Shorty's Not So Lost Mines and Treasures of Southern Oregon

By Steve "Shorty" Owen
2005 © S.W. Owen

ISBN # 1442176202
EAN-139781442176201

Executive Editor: Robert Owen
Development & Production: Steven Owen
Formatting Editor and Design: Steve Owen

All Photo's by Steve "Shorty" Owen

Dedicated to my son
"Adam Christopher Owen"
1975 – 1995

In addition, thanks to all those who helped put this together.
Linda, Alex, Robert, Brenda, Doug, and Larry.

Also, a Thank You to my dear friend
"Miner Don" McPherson
He passed away in 2006 and is soulfully missed.

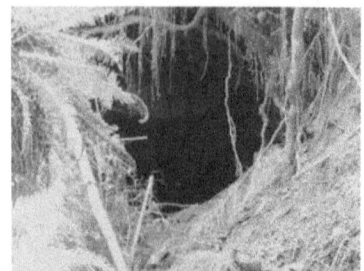

The Umpqua Mine

1897 Miners Cabin

Aluminum Mine up Doe Creek

The Copper Mine

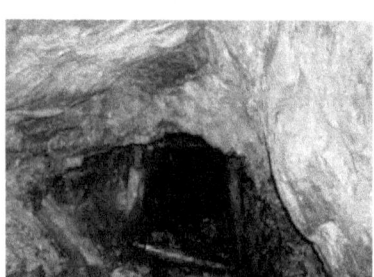

The Chi Do Mine

The First Strike Mine

The Ladder Shaft Mine

Miner Don in front of the First Strike Mine.

Mines and Treasures of Southern Oregon

Table of Contents: Page

Places to see while in

Canyonville, Oregon

Here are some great places to go while you're here

1) **Canyon Café** a great place to eat Breakfast or Lunch. Located at the corner of Main St. and 5th St.
2) **Bead Mecca** = Beads and Beading supplies. Gallery of Gifts 425 S. Main St, http://www.beadmecca.com
3) **Mara Kay's Cafe** = home style cooking at its best Located at 412 S. Main St.
4) **Monkey Business**= Guns and Tools and great bargains too. Located at 325 S. Main St.
5) **Ken's Sidewalk Café** a pleasurable place for a milk shake and lunch, the best and biggest burgers, fries, onion rings in town. Located on the corner of Main St. and Canyonville / Riddle Rd.
6) **Canyon Market and Liquor Store** a great place for supplies for the day Located at 425 N. Main St.
7) **Power Pit Gym** buff up or just work out a great place to make new friends Located across the street from Canyon Market.
8) **Promise Natural Foods & Bakery** = Full Natural Food Grocer & Whole Grain Bakery, Organically Grown Produce. Located at 503 S. Main St.
9) **B & C Antiques** = Stop in and look at the Largest Antique Store in south county. Located at 431 S. Main St.
10) **Foresters Lounge = Spirits and Food A great Gathering Place**. Located at Next door to Canyon Café.

Local History
South Umpqua Historical Society Museum
421 W. 5th St.   541-839-4845

A Place of Fun for the Whole Family

Notice to all

No mine is ever safe to enter.

Please use your own judgment.

If you are compelled to enter any of these mines please leave someone outside to go for help.

AND TAKE ALL PRECAUTIONS TO INSURE A SAFE DAY!

I take no responsibility for your actions!

PLEASE LEAVE THESE SIGHTS AS YOU FIND THEM!

These are either Historical sites or **VERY** important memories.

I hope that they will be kept in a manner that will allow all to see and enjoy for many years to come.

I am working on getting them all listed as historical sites.

If this doesn't happen I am going to ask for help in cleaning them up and making them accessible to all.

My thoughts are, if some would adopt a mine (just like when you adopt an area of the freeway for clean up) we can make these sites remembered for all time.

They are a part of our history. If we don't keep them alive our grandkids won't know anything about our past.

In addition I need your help to stop the government and special interest groups in closing all the "Public Lands" as they call them. If they had their way we would never be allowed to go out in the forests at all.

We need to as a work group and make lots of noise! Remember, The loudest voice get the most attention!

There is no information on this mine but it is dangerous inside. It was fun going deep inside (about 400 feet in the main tunnel and 3 adits each about 175 feet) but I do not recommend doing it.

Oregon Opal found near Tiller Oregon.
Pieces were with-in three feet of one another.

A Bit Of History for Southern Oregon

It is known that Gold was first discovered in Oregon in the summer of 1850 by a group of miners. Most of which were from Illinois. They worked a placer known as the "Illinois River" near the mouth of Josephine Creek.

This discovery is responsible for starting the Oregon gold rush in southwestern Oregon in December 1851. The rush for gold began near what is now known as Jacksonville, Ore.

Jacksonville was a campsite on the overland supply route to the California gold fields. Thousands of men came into Jacksonville mining camp to stake claims and make their fortunes.

Word got out about the lode and placer mines in the Southern Oregon area and Jackson County. It soon became the *hub* for all miners coming into Oregon.

At one time during the gold rush, Jackson County was the most populated county in the state.

Mining camps sprang up overnight in such places as Jacksonville, Buncom, Sailors Diggings (now known as Waldo, Kerbyville (Kerby), Willow Springs, Phoenix, Allentown, Browntown, Golden, and Placer.

The richest placers in the area were found on Sterling Creek, Althouse Creek, Sailors Diggings, Rich Gulch at Jacksonville, and Rich Gulch at Galice.

Most of the major placer producers were located on Sucker Creek, Josephine Creek, Graves Creek, Foots Creek, Briggs Creek, Galice Creek, Sardine Creek, Galls Creek, Forest Creek, Ferris Gulch, Powell Creek, Poorman Creek, Palmer Creek, and Powell Creek.

The Chinese played a very big part in the building of ditches and levies for diverting water. Hydraulic mining was the preferred choice of the times. The most famous of the hydraulic mines was the Sterling Mine, which had a twenty-three mile ditch, fed by the "Little Applegate River".

Built in 1877, and some of the ditch and equipment is still visible today. Lode-gold mining got big in 1880. The biggest producer at the time was the Greenback mine. It was discovered in 1897. By 1939 it had produced more than 3½ million dollars in gold!

Between 1900 and 1910 the Greenback employed more than 100 people. By 1916 it employed only 30.

Gold production in Oregon is hard to calculate figures on as not all gold reached the U.S. Mint.

Gold, Platinum, Silver, and Copper were used in daily trade and were shipped out of state for payments of all manner of goods. However, in 1914 the reports from the U.S. Mint showed receipts of gold being sold between 1851 and 1882 was about $17.2 million. More than half of this was believed to have come from southern Oregon.

In 1852 gold was discovered on some Oregon Beaches. And, for a time some proved to be very profitable.

A few years later gold was found on most of the ancient elevated beaches. The earliest beach gold was found at the mouth of Whiskey Run a few miles south of the Coquille River. This area was found to be rich in ore.

The gold ore was what everyone was looking for. But, a few in 1927 were smart enough to find platinum.

Platinum was as rich in the beach sand as gold. The only problem prospectors had, was removing the micro size platinum and gold particles from the sands.

Elaborate separators were needed to separate the ores. Men from all over the state tried everything.

Mercury was the only way they really new of that would work. A man named Collegin Walters came up with a flat table that he called a rocker table or separator table.

Water flowed over small ridges on the table. The sand is then poured on the table allowing the water to wash through it. This would separate the gold and platinum causing it to flow to the end of the table and fall into a small bucket.

Prospectors explored essentially all of the streams in Douglas County for placer gold. They found fair to good prospects on only a few of the streams. (Noteworthy among these are Starvout, Hogum, Quines, Bull Run, Coffee, and North Myrtle Creeks).

Other streams on which placer mining was done included Steamboat, Windy, Cow, West Fork Cow, Beaver, Jordan, Willis, Drew, Coarse Gold, Byron, Thompson, and Bushnell Creeks.

The early-day placer mining was done mainly by hand. Usually by shoveling into a rocker box or long tom sluice box.

In the richer streams miners were often able to stake out only a small plot and had to work them any way they could.

If they bought a sizeable area and had enough financing a ditch and pipe were installed so that the area could be mined by hydraulic mining.

December 18, 1858 M.J. Suark wrote a letter printed in The Oregonian newspaper (1939) by Spreen:

"Coffee Creek is a small branch of the South Umpqua River. Three men had been working the mouth of the creek for several months with sluices, but owing to the fineness of the gold and bad management, very little was made by them. In the latter part of July 1858, a company of men, including one of the three, started up the creek to prospect. They worked until September when they struck a prospect three mile from the mouth that paid them four dollars per day per man.

Further prospecting located a strike that paid as high as two and a half dollars per pan per man a day. Each of the party, consisting of seven men, staked claims of two hundred yards each. All the prospectors on Texas Gulch a tributary of Coffee Creek was getting as much as two and a half dollars per pan on bed rock. One man took out a nugget weighing more than six ounces. The Bonanza quicksilver deposit was discovered east of Sutherlin some time during the 1860's. The Gold Bluff mine southwest of Canyonville was discovered in the late 1890's and the Silver Peak mine in 1910.

Large deposits of nickel at "Nickel Mountain" west of Riddle (also known at that time as Old Penny Mountain) was discovered by sheepherders in 1865.

The green ore (garnierite) was thought to be both copper and tin before it was determined by assay to contain 6 percent nickel and no copper or tin.

An 1898 report stated that placer mining on Lee Creek and Buckfork Creek of upper North Myrtle Creek had been active during the rainy seasons for many years.

Production prior to 1898 was estimated to have been about $150,000. The Chieftain-Continental vein on South Myrtle Creek was discovered in 1898, shortly after Diller mapped the area geologically.

The story of placer mining activity on upper Cow Creek is not well recorded. It has been reported that in the early days two placer camps were working on creeks now called Starveout and its tributary, Hogum Creek.

Occupants at the camp on Starveout Creek ran out of food and visited the camp on Hogum Creek to see if they could purchase supplies with newly-mined gold, but they were refused. The result was some hard feelings, name-calling and subsequent naming of the creeks, a tributary of Hogum Creek, called Fizzleout tells a story of its own.

Quines Creek and its tributaries, Bull Run and Tennessee Gulch, had fairly good placers and were centers of considerable activity. Probably during the late 1850's and 1860's.

A Chinese settlement is said to have been established near the head of Tennessee Gulch.

General Geology

Within Douglas County are five distinctive geomorphic provinces in western Oregon. These include the Klamath Mountains, The Western Cascades, the High Cascades, the Applegate area, and the Coast Range.

Characterized by a more or less unique group of rocks. This in turn is responsible for the particular topographic expression and mineral resource of the whole area

The oldest rocks in the county are from the Mesozoic age. They are restricted to the very rugged area around the Klamath Mountain range in the southern portion of Douglas County. Consisting primarily of marine sediments and volcanic rocks, having a composite thickness of about 6 miles. Tertiary rocks underlying the more gently deformed Coast Range area consist of a composite thickness of 10,000 to 20,000 feet of submarine basalt and rhythmically bedded sandstone and siltstones.

Klamath Mountain Areas:

The Klamath Mountains are an area of complex geology and extremely rugged topography. The leading major streams include Cow Creek and other tributaries of the South Umpqua River. Cultivation is only allowed on the lowlands that border the rivers and streams. Mining in this area has produced a variety of very profitable and useful metals including gold, silver, copper, chromites, nickel, mercury, and zinc.

Industrial minerals and potential non-metallic ores in the areas include sand, gravel and crushed rock, building stone, barite, limestone, asbestos, talc, olivine, rubies, garnets, and many other semiprecious gem rock.

Applegate Area

Walls and others in 1949 defined the Applegate area for exposures south of the county line in the Applegate River Drainage. The Applegate area consists of several tiers of Slate thousands of feet high. Slate siltstones, Chlorite Schist, Quartz-Schist, and Quartzite representing meta-mophosed Sandstone, Shale, and huge deposits of volcanic rocks and possibly in part from gabbros related to the ultra-mafic rocks associated with the area.

The Applegate area is approximately 30 square miles of terrain in south central Douglas County and extends southward into Jackson County. This area is intimately loaded with serpentines and peridotites and is intruded by quartz diorite of Mesozoic age.

Metal ferrous mineral deposits occurring in the Applegate Area include gold, silver, copper, zinc, mercury, and manganese. Potential industrial mineral resources from the formation include talc, mica, garnet, and building stone, mica, garnet, and building stone.

Coast Range Province

The Coast range Province consists of Tertiary submarine lavas and marine sediments. The region is less deformed than the Klamath Mountains Province and exhibits lower relief. Nonetheless, the terrain is rugged in places owing to the resistance to erosion of many of the sandstone interbeds and small intrusive bodies.

Major peaks include Bear Mountain (elevation 3,180 ft.), Old Blue (elevation 2,536 ft.), and Roman Nose Mountain (elevation 2,860 ft.). Mineral wealth of the area is limited to a few unworked deposits of impure coal, localized quarry rock, with sand and gravel deposits.

Western Cascades Province

The Western Cascades Province is known as a rugged topography having irregular ridges and deep narrow valleys. The region is underlain by a variety of late Eocene through lat Miocene volcanic and subordinate fluvial sedimentary rocks. The higher peaks range upward to 4,000 or 5,000 feet in elevation.

Small intrusions are scattered throughout the area and localized mineral deposits include cinnabar, antimony, gold, platinum, silver, copper, and silica.

High Cascades Province

The High Cascades Province is known by its high plateaus of Pliocene and Pleistocene volcanic rock capped by a variety of volcanic peaks, which include Mt. Thiesen (elevation 9,182 feet), Mt. Bailey (8,363 feet), Trap Mountain, Cowhorn Mountain, and Elephant Mountain. The rocks are essentially flat lying except for initial dips and are among the least deformed strata in the county.

Rocks of the High Cascades area are younger then the rocks in the Western Cascades province. They range in age from Pliocene through the Pleistocene and include a variety of flow rocks and air-fall pumice deposits. The bulk of the Plio-Pleistocene flows (Cascade Formation) consists of open-textured olivine basalt and olivine-bearing andesite and ore designated on the geologic map.

Less extensive extrusions of late Pleistocene light to dark-gray, dense to open textured olivine basalt and olivine-bearing andesite form intra-canyon flows north and south of Pig Iron Mountain. Westerly remnants of the flows are preserved as several small patches in the valley of the North Umpqua River in the Western Cascades Province.

In the headwaters of the Rouge River in the extreme southeastern part of the county deposits of white to buff dacitic ash fall and ash-flow tuff form rudely sorted valley bottom deposits in the valley of the Rouge River drainage. Eruption of these ash deposits from Mt. Mazama occurred between 5,000 and 7,000 years ago.

Mineral resources of the High Cascades area include pumice, cinders, and lava rock of various compositions, most of which makes excellent road aggregate. Emery is reported from near Pig Iron Mountain.

Pictured is a sliver of Cellulite Rock off a wall about 10 feet tall by 5 feet wide. About 10 miles from Jackson Creek, near Tiller.

Hydraulic Mining on North Myrtle Creek in the early 1900's

(Photograph courtesy of the Douglas County Museum)

Covered Bridges are all over the Southern Oregon Area

NOTES

Gold Mines and Occurrences by 1937

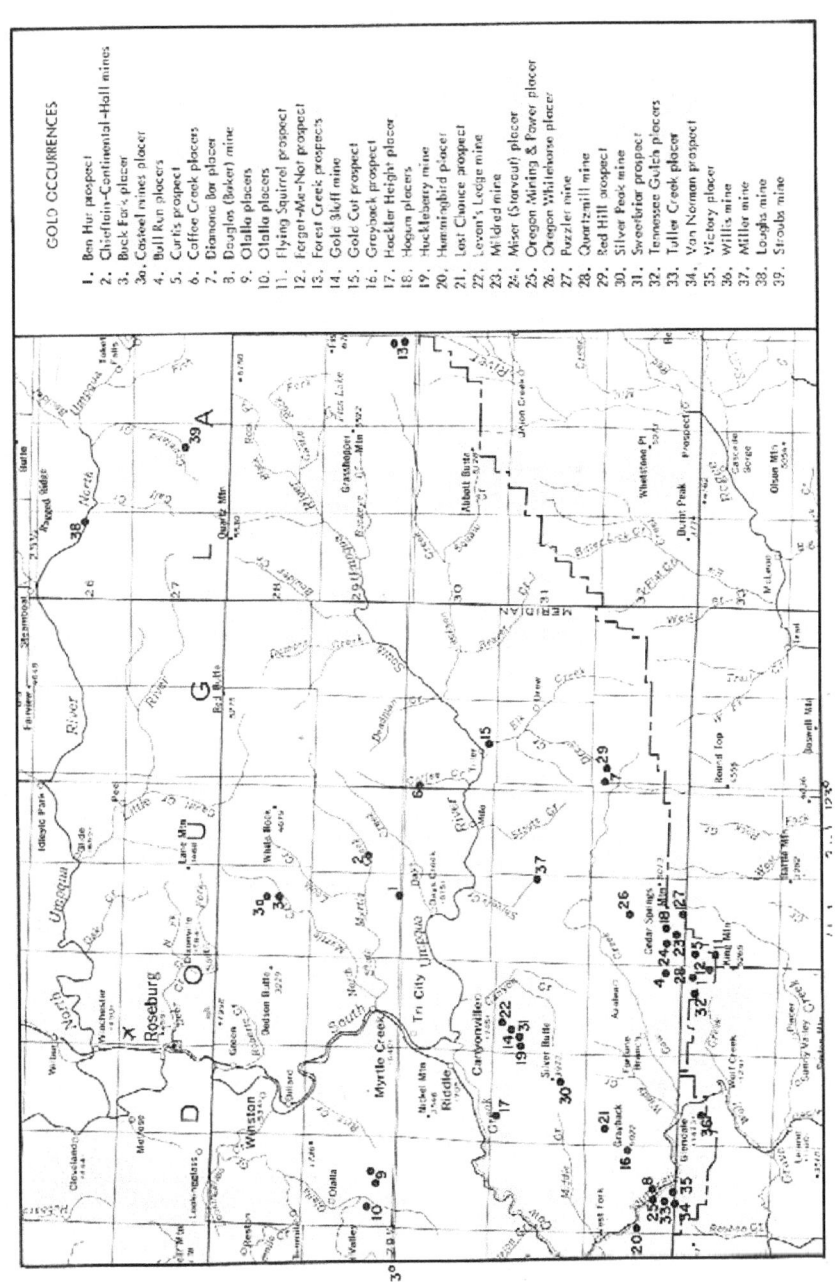

GOLD OCCURRENCES

1. Ben Hur prospect
2. Chieftain-Continental-Hall mines
3. Buck Fork placer
3a. Casteel mines placer
4. Bull Run placers
5. Curtis prospect
6. Coffee Creek placers
7. Diamond Bar placer
8. Douglas (Baker) mine
9. Olalla placers
10. Olalla placers
11. Flying Squirrel prospect
12. Forget-Me-Not prospect
14. Forest Creek prospect
15. Gold Bluff mine
16. Gold Cut prospect
17. Greyback prospect
18. Hockler Height placer
18. Hogum placers
19. Huckleberry mine
20. Hummingbird placer
21. Lost Chance prospect
22. Leven's Ledge mine
23. Wildcat mine
24. Miser (Starvault) placer
25. Oregon Mining & Power placer
26. Oregon Whitehorse placer
27. Puzzler mine
28. Quartzmill mine
29. Red Hill prospect
30. Silver Peak mine
31. Sweetbriar prospect
32. Tennessee Gulch placers
33. Tuttle Creek placers
34. Van Norman prospect
35. Victory placer
36. Willis mine
37. Miller mine
38. Laughs mine
39. Stroubs mine

21

The Old William O'Macklin
Or
"The Old Willie Mine"
Opened in 1876

N.42°45'02.8"
W123°13'09.8"
Elevation 2319 Ft.

The Old William O'Macklin or "The Old Willie Mine"

William O'Macklin found Gold and Gems making him a very rich man; he died at the age of 63. The mine is in OK shape. Partially maintained through the years have kept it a beautiful part of our history. New timbers were put in place in 1947.

Not much more information can be found about this mine. It is very old and much information has been lost.

The Old Willie is a great place to start your treasure-hunting day.

If you enter this one be careful!

The surrounding hillside is not very stable. Observe the mountain and terrain surrounding and you will see what I mean.

Enter at your own risk!

Directions for those who do not have a GPS

Interstate 5, take exit 86 at end of off ramp go East, to stop sign, turn left, turn right on Quines Creek. Go another 1-½ miles to Bull Run, turn left. Follow this road about two miles the road will fork, stay right, go another 1 mile, and watch the left side of the hill. You will drive right next to the opening.

If you continue on the same road you will find yourself in about 20 miles back on a paved road and able to get right back on the freeway. This road and a few that junctions off it will take you to about 30 mines!

The "Umpqua Mine"

Opened in 1872

N. 43°01'03.8"

W.122°55'58.5"

Elevation: 1928 ft.

The "Umpqua Mine"

The Umpqua Mine is very easy to find and a great one to look at.

As you can see by the picture the opening has been blocked up to keep people out (Not me of course).

The mine at the time of this photo (March 5th 2005, about 1:45 PM) was about 4 feet deep in water from the snow and rain run off.

When this mine dries out some it is full of color! It is about 190 feet deep. This one is a real treat to look around inside.

There are many obstacles so **be very careful** when checking this one out. Although this mine is not that old it does have a great history. There are many pages on the Internet that will tell you more than I can here.

The shaft has an opening of about three feet round. It then opens up quite nicely within a few feet.

I would do my best to check this one out while your in this area.

Be sure to pack a lunch!

If you head back to the main road and go east you will find "Umpqua Falls Camp Grounds." A great place to spend your lunchtime.

(The old Jackson Creek Store should have all the supplies you will need for you picnic lunch or dinner).

Some very nice people live in this area so if you have problems do not fear asking for help!

Spend the day sight seeing, and treasure hunting. Gold, Gems, Silver, Platinum, exists in almost every creek, river, and mountain in this area!

The "Turkey Creek#1"

N42°52'39.5"

W123°14'28.1"

Elevation: 1167'

The "Turkey Creek #1"

The Turkey Creek #1 Mine is a comical hole. It is so close to the highway more than 1,000,000 people drive by it every year, yet, no one knows its there! It's only 100 feet off Interstate 5.

It is a very beautiful mine, about 90 feet long and hidden to all except a chosen few and now you.

Its history is very tarnished at best. 15 Chinese workers were killed while digging this one. The shaft collapsed when they were at about the 75-foot marker. The mine collapsed with almost all the men inside working, leaving too few men to dig them out.

The men that were there tried they're best, but it was too much for them to do alone. The ones still inside didn't live long for the thick dust and very little oxygen.

I figure that they lasted about 12 minutes.

The men that were left continued to dig until they reopened the mine and removed the dead.

To this day men still go into this mine to look for riches. A few come out with the spoils of a good days work. There is still gold in this mine but there is also "**Mercury**"!

Keep your gloves on in case you come across a pocket.

This can be very dangerous, and it smells bad!

The Quartz Vein

I just found this spot today while out tracking down some other old mines. I came around a corner and there it was! This vein is at least ½ mile long. The vein ranged in size from a few inches to more than five feet wide! I took a very large piece home to see.

The vein is up a very nice dirt road not more than 3½ miles from Tiller Trail Hwy. There is so much to see here. Many different rock formations, eagles, and all sorts of wild life, Garnets. **Yes, I said garnets!**

This area is full of Garnets from "BB" size to very nice ring and necklace sizes. You will have to see it for yourself to really understand its beauty. **Remember, the area is wild and full of wild life**! There are deer, elk, bear, cougar, skunk, and (just to keep it interesting) rattlesnake.

Always be prepared for most anything! Viewing the sunset from the 3,000' elevation is simply breathtaking. I love this area!

Most of the trees have been removed for timber. But, that actually makes the view (written with mixed feelings) better. The removed trees have actually helped our mission getting to the gold and precious minerals. Don't worry I'm sure the area will soon be re-forested.

This area is a major gold mining area from the late 1800's. At one time Coffee Creek had a population of more than 2200! People that is.

Over 200 tons of gold was mined from the Coffee Creek basin alone. In addition, Coffee Creek leads you to the "Texas Gulch" area.

Texas Gulch is a story in itself.

GPS coordinates to the Quartz vein:

N43°03.1616' W122°46.6047' Elevation: 998.9ft

The Legend of the

"LEDGE OF GOLD"

The "Ledge of Gold' is a story that has been told for over 25 years.

Rex, a retired logger was working on the northwest side of Canyon Mountain when his bulldozer cut around a corner and he saw a line of yellow.

It is said to be more than 25 feet long and more than eight inches thick. He stopped his bulldozer to take a closer look. When he got up close to the line of yellow, he knew it was gold! Rex then went back to the bulldozer to get some tools to take a little piece back home.

The next day he took the sample to the local gold exchange store in Roseburg. The exchange told him it was very high grade Gold and asked where he found it.

He told them he found it out side of Riddle. (about ten miles from where he really found it).

Rex went to work on the same road the next day. But, this time he did some extra work. He decided to change the route of the road he was building to go around the other side of the hill from the original plans.

He then piled up some dirt to make it look like no one had tried to make a road that way. His thought was, if, after the logging was done and everyone moved out of the area he would go back in and take more Gold out. Well, sad to say, this never happened. Rex died that winter of pneumonia.

He never told anyone where the ledge was. I live just a few miles from the area and can see it very well from my front yard. I cannot seem to see where he could have found it, but it is fun to look for it.

Rex is remembered today for all his hard work during the glory days of logging in the northwest. Some of the equipment used then would turn the stomach of most of the loggers today.

Rex's family still thinks about where the ledge could be. But, he hid it so well that until now with today's new technology is it maybe possible to find it.

Pictured is my son walking to what I think is the Lost Sailors Mine. On Cow Creek Rd. Mile marker 8.

NOTES

The "Iron Door Mine"

N42°53'40.3"
W123°20'08.1"
Elevation: 2225 FT.

The Iron Door Mine

This is a very old and not very well kept mine; including the road getting there. It was littered with trees that had been blown down in recent storms. We had to move more than 25 large trees. We just drove over the little ones.

This one is a great find with lots of color. There are a lot of green slimy looking rocks. They feel oily to the touch.

It is loaded with nickel. The mine goes into the mountain about 20 feet. Beyond that it looks like it has collapsed. The shaft should have been more than 200 feet long!

When you enter this one, look to your left and you will see a small opening. When you climb up to get a closer look you will see a small hole. Shine your flashlight into it and you will see down the shaft for at least 50 feet.

It was a very productive mine in its day and very well could be again. You will need a four-wheel drive to get to this one.

Side-note: There is another mine within 250 feet of this one. The "Yellow Jacket 1".

The "Yellow Jacket 2" is supposed to be within 100 feet of #1.

.

The Yellow Jacket #1

N.42°53'47.4"
W.123°20'04.5"
Elevation: 2038 FT.

The Yellow Jacket #1

This one goes straight down, there is a fence up around it so people don't just fall in. At the time of the photo it was filled with water.

Probably 20 feet deep.

It belonged to my wife's Uncle and now belongs to me.

I know a lot about this one. This hole in came into existence in 1934. He worked it for nearly 20 years before moving to one of 9 other mines that he left Linda and I.

He did go back from time to time and keep his assessment current. Since he died it has been left alone (except some work I have done around it).

I was never able to gather from him where the Yellow Jacket #2 is located. But, I do know it is between 100 and 250 feet from #1.

Driving directions:

From I-5 take exit 98 or 99, go into Canyonville to Canyonville School. Turn towards Riddle, go another 2 ½ miles to Shoestring Rd. Turn left. Go about 1 ½ miles, you will see a BLM road number 31.0-6-35.0 on your left.

Turn left, go ½ mile the road forks stay right and follow up the hill. Go another 1 mile and the road forks again stay right. At the next fork go left road number 31.0-6-12.1, this is where the 4X4 will come in handy. Go 1-¾ miles and follow the GPS readings. The Mine is in a valley on your left. Have fun and be very careful.

Its only about 100 feet, follow the road on the other side of the wash. When you come to the main road again turn right and look to the right. You will see a big cut out of the hill, at the bottom of the draw you will be at the "Yellow Jack #1."

The Donna May

N. 42° 43'59.5"

W.123° 12'04.5"

Elevation: 3151 Ft.

The Donna May

The Donna May is one of the mines that you can drive right up to the front door. It is about 12 feet off the road.

Opened in 1923, it was given to my lovely wife's Uncle in 1979.

The Shaft is approximately 100 feet long. It is full of color and very stable. Do not fear going into this one. It now belongs to Linda's cousin "Donna May."

Before Linda's uncle took over the mine it was called the "Horseback Mine". Before that it was called the "Uncle Bill".

I do know that a Chinese rail worker opened it in 1923. He only worked it for a few years and went back to China. No record was made as to the name of it at that time.

The mine produced copper, silver, some gold, and a few gems while Linda's Uncle owned it. His daughter, Donna May, wanted to keep it in the family as it was the mine her dad was working at the time of his death. This area is full of bright shiny green and blue rocks symbolizing the presence of Copper, Nickel, Iron, and Gold. Many men have found they're fortunes in this area.

The Quartz Mill Mountain has long hid its full treasures.

Even today many of the local miners and prospectors search for the hidden treasures of Quartz Mill.There are some 45 mines and claims around the mountain, make sure when you check this one out you watch for the signs of an open claim.

Driving directions:

Interstate 5 to exit 86 go east to stop sign, turn left, go just past restaurant and turn right on Quines Creek Rd. Go one mile to Bull Run, turn left. Follow Bull Run (it will make a few turns just stay on the main road). You will go past a few houses. **Please follow the speed limit signs.**

You will drive right by the "Old Willy " mine at the 1-½ mile marker. At the "Y" stay right. Follow Bull Run (it will turn into a numbered only BLM road).Follow the main line up.

The road will fork again at the 2-1/2 mile marker. Go left.

One more mile and the Donna May mine will be on your right.

The Shoestring Mine
Sad to say this one is on Private property now. But boy is it fun.

Mercury Mines of 1946

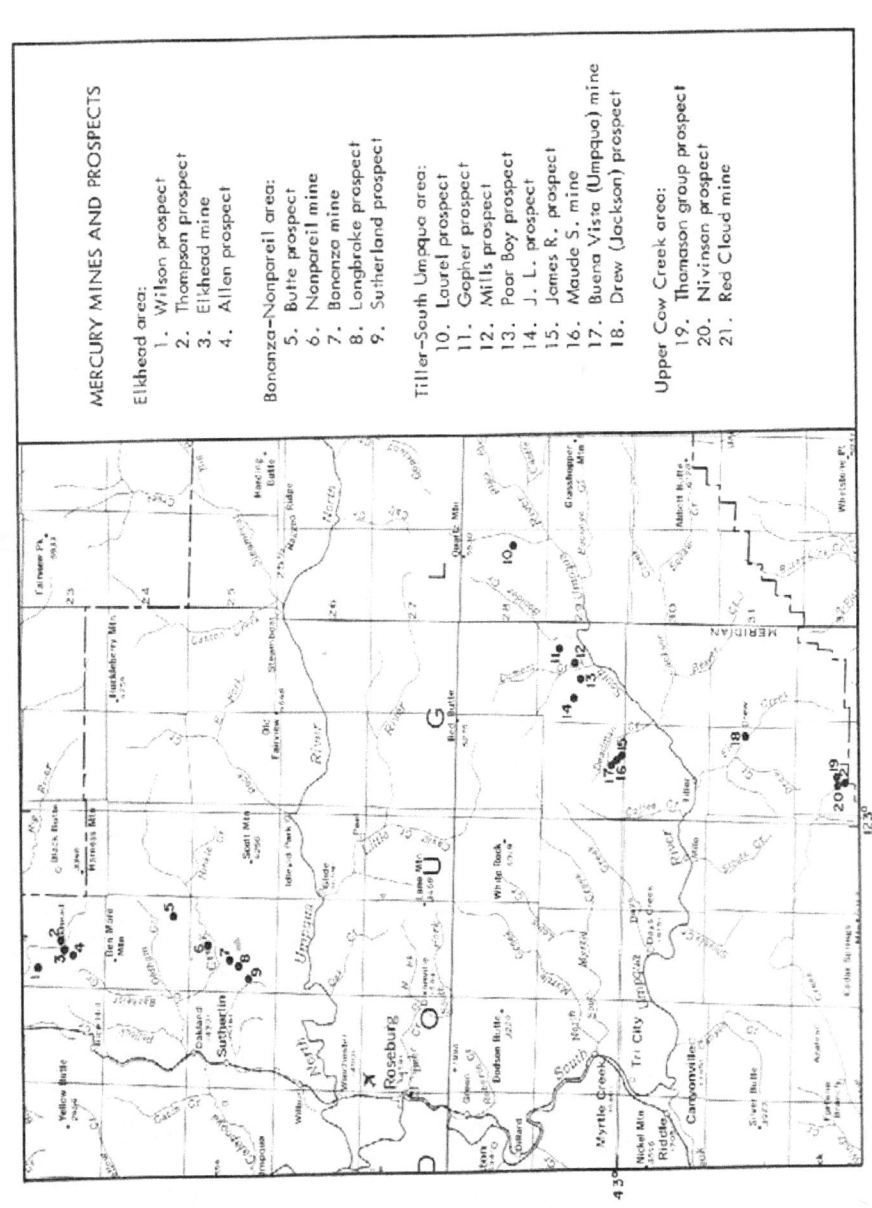

MERCURY MINES AND PROSPECTS

Elkhead area:
1. Wilson prospect
2. Thompson prospect
3. Elkhead mine
4. Allen prospect

Bonanza-Nonpareil area:
5. Butte prospect
6. Nonpareil mine
7. Bonanza mine
8. Longbroke prospect
9. Sutherland prospect

Tiller-South Umpqua area:
10. Laurel prospect
11. Gopher prospect
12. Mills prospect
13. Poor Boy prospect
14. J. L. prospect
15. James R. prospect
16. Maude S. mine
17. Buena Vista (Umpqua) mine
18. Drew (Jackson) prospect

Upper Cow Creek area:
19. Thomason group prospect
20. Nivinson prospect
21. Red Cloud mine

The "Copper Mine"

N. 42°43'59.0"

W. 123°12'17.9"

Elevation: 3755 FT.

The "Copper Mine"

The Copper Mine is a hard find. It is hidden along side a wall of rock. It goes almost straight down about 20' and looks like it has collapsed in a few places.

There is not much I can find out about this one. It was opened in 1912, and three lost their lives trying to get it secure and safe. It was reopened in 1956, again in 1983. It is a closed claim at the time of this writing.

This area as it is known to produce some very fine gems. High quality Garnets, Rubies, and Agates are scattered over this region. There might be one under your foot at any time!

If you look around the area keep an eye open for a mine that has been lost for many years. It is said to be within ¼ mile of the Copper and L and M Copper Mines. The Ruby Mine was lost to a cave-in and land slide (1919).

If you look just down the hill from the Copper or L and M Copper you can see where a slide had happened a long time ago.
Good luck, with your searches.

Driving directions: Follow the direction to the Donna May, when I directed you to take the second "Y" to the left to the Donna May. All you have to do is turn right and follow that road all the way until it dead ends.

You will find yourself at what is called a "Mill Site". This is where two or more mines processed they're materials. They would all share the equipment, and share in the expense of repairs.

From the mill site, follow the old walk and cat trail up the hill, its about 200 yards to both the "Copper" and the "L and M Copper" mines.

The Lost Treasure of

"General Yau Pin Auop"

The treasure is said to be in an area between Glendale and Riddle. Some think it is near the old Railroad tunnel that the Chinese built. This area around the tunnel is called "Peck." After the day's work on the railroad was done, they would go to work digging for gold. A vein was found behind the tunnel so they would sneak in after work and peck at the walls.

The Railroad soon went in and had them pour concrete to build walls. Therefore, they would stop digging around the tunnel. This did not work for the Chinese workers just dug their own tunnel behind the railroads tunnel. It is suspected they found the vein, and removed more than 3000 lbs of the yellow material.

In addition, while up there be sure to keep your eyes open for the infamous lost treasure of "General Yau Pin Auop." It is told that while working with the railroad workers, he would keep part of their finds as his share. He buried the shares of gold from the workers somewhere in this area.

If you take I-5 to exit 103 and go west, about 16 miles you will see the tunnel for the railroad on the right side of Cow Creek. This is by far one of the best gold producing creeks in Douglas County. At the 21-mile marker, you will find a county park for prospecting. You can pan, dig, and use a sluice box. NO motor operated dredges allowed. This is a great trip with allot to see and do. Allow about three hours to take this trip. It is a 36-mile trip, with places to camp, picnic, and pan for gold.

Pictured above is China Camp on the left, more than 150 acres where more than 10,000 Chinese, Irish, and Germans worked the railroad. During the day and panned and mined at night,

On the right is some of the two pounds of coins I have found at China Camp with one of my metal detectors. The oldest coin is 1876 and the rest vary from 1889 to 1955. One of the oldest is the 1892 Chinese coin, (it's the one with the square hole in it.

The Lost Sailors Mine

This is a story about three sailors that had come to shore in Crescent City California in 1867. They had one year to travel the US and see the sites before meeting back up with their ship in Astoria, Oregon.

They decided to travel through Oregon and up to Glendale and do some prospecting. They bought all their supplies at the local mercantile. They set out to travel up Cow Creek towards Roseburg.

Now according to all stories they traveled about 21 miles up Cow Creek, and spent a week in a small cove. While two of the sailors were working the creek, the other thought he would try his luck at the hill. After three days, he came to camp one evening and showed the others what he had found.

They were soon all digging in the hill. They had found more than 25 Lbs of gold from this spot in less than a month. The three men decided they would travel to Roseburg and get a claim on this spot. When they arrived, they were told that no one that was not a resident of Oregon could get a claim in their own names. They spent two weeks in Roseburg trying to figure out what to do.

They soon made friends with a local rancher, who told them he would get the claim in his name. Therefore, when they got out of the Navy, they could come back, work it, and become residents of Oregon. The rancher promised them he would never tell anyone where the mine was, and would never work the mine until they came back.

The rancher was true to his word. Problem was the rancher died two years later. He never told anyone where the mine was. The sailors were killed in battle at sea, never to return to Oregon and their mine. I have narrowed the search down to a two-mile section of Cow Creek. I feel this has to be the place where the "Lost Sailors Mine" is. I have spent five years trying to retrace the flow of the creek since then. Floods and mountain run-off have changed the look of the creek.

Indian Ceremonial Grounds

This area is being kept a secret by me. I found this area while out searching out an old mine shaft.

There are more than 30 fire pits and two "Infinity Fire Pits."

There is also a large rock wall with four fire pits below it and groves from where the "Holy Man" would have stood.

The stone is about six feet tall with a shear face on the north side of it. You can see where the footprints have smoothed out groves in the stone.

These are just a couple pictures of the site, one is of the "Infinity Pit" another is of about fifteen fire pits, and the last is of a large "Council pit".

Above is a picture of the Infinity Pit, below is the Council Pit. The picture at the bottom is of the area with more than 15 pits in one section.

NOTES

The "L and M Copper Mine"

N. 42° 43'58.2"

W. 123° 12'18.3"

Elevation: 3761 FT.

The " L and M Copper Mine"

The L and M Copper mine isn't very old. It opened in 1937. The price of copper was rising due to the needs of the U.S. military at the beginning of World War II.

This mine is only about 40 feet deep, but is so full of color its unreal. If you head for this one be sure to take a flashlight and a piece of red plastic to cover the lens. When you get inside and your eyes adjust with just the flashlight. Cover the lens with the red plastic, get ready you wont believe your eyes.

The explosion of color will almost take your breath away. The mixture of green, blue, white, reds, and purples are fantastic. Yes, I did use the colors as a plural. The many shades and styles of colors are incredible.

You have to check this one out. It's not that deep but you will have no problem spending a couple of hours looking around this area.

Directions: From I-5 take exit 86 go east to stop sign and turn left, go just past restaurant and turn right on Quines Creek Rd. go about one and a half miles to Bull Run Rd. turn left, follow Bull Run about 2 miles you will pass a few houses please go slow and watch out for the crossing signs. You will see a DEER crossing, CAT crossing, and GEESE crossing signs.

The Old Blacky Mine

N. 42° 44'16.2"

W. 123° 11'16.3"

Elevation: 3142 FT.

The Old Blacky Mine

This is a very old mine. It is across a creek and up the side of a lightly sloped hill. It has not been well maintained and has a lot of brush and other debris in front of it.

The mine was opened in 1874. No record of the owner's name, only the name of the mine itself.

It took me more than an hour to track the tailings piles to the shaft.

The mine did produce a lot of gold. Totaling more than 300 lbs. in a 15-year period. With that amount of gold produced you'd think the owners name would be found somewhere.

Driving directions:

Interstate 5 to exit 86. Head east to stop sign. Turn left. Turn right on Quines Creek Rd. Go one mile to Bull Run. Turn left. Follow Bull Run it will make a few turns just stay on the main road.

You will go past a few houses. (**Please follow the speed limit signs**) you will drive right next to the "Old Willy " mine at the 1-½ mile marker. The road will "Y", stay right. Stay on Bull Run. It will turn into a BLM road and be numbered only. Follow the main line up. The road will fork again at the 2-1/2 mile marker. Go left. Just past the Donna May the next small valley on the right. Stop and search the eastern side of the slope.

The Whitehorse Mine #1

N. 42° 47'41.0"

W. 123° 09'65.2"

Elevation: 1745 FT.

The Whitehorse Mine #1

Not much can be found about this mine. It was opened in 1895. The owner is buried somewhere inside the original shaft which is actually twenty feet south of the opening now. It looks like someone tried to go in and meet the main shaft but failed.

The new shaft only goes in about 15 feet. It is stable and looks like a bobcat has been using it for a den. There is old half eaten bones of a small animal. There were no signs of any recent visitors.

This mine cannot be seen from the road. Yet you park right below it. There is a camping area about 190 feet away. From the mine you can see the campground but cannot see the mine from the campground.

When you arrive at this one, take a big breath, and start climbing the hill on the east side of the road. The campground is on the west side.

About 160 feet up the side of the hill you can see a deer trail, follow it. You will see the mine entrance behind a pile of stone and dirt. If the mine has an old spring bed in front of it, that is the wrong mine. I will tell you about that one next. If that is the one you found first, follow the deer trail south at the same elevation you will come to it about 75 feet away.

Directions: Take exit 188 off I-5 and go towards Galesville Dam, go about 4-1/2 miles to Longfibre Rd., turn right, follow Longfibre about 2 ½ miles to a campground on your right. Park here and walk the rest of the way.

Copper found in Douglas County 1946

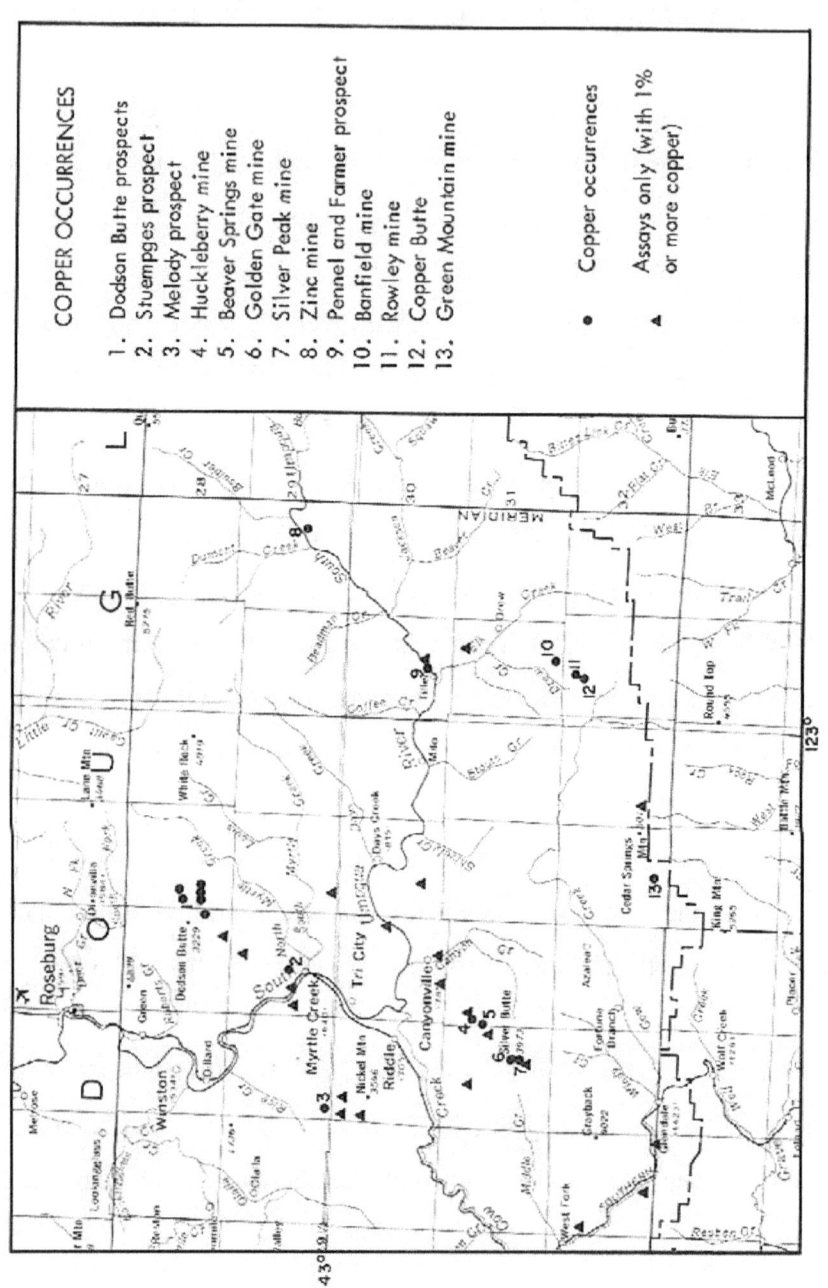

COPPER OCCURRENCES

1. Dodson Butte prospects
2. Stuempges prospect
3. Melody prospect
4. Huckleberry mine
5. Beaver Springs mine
6. Golden Gate mine
7. Silver Peak mine
8. Zinc mine
9. Pennel and Farmer prospect
10. Banfield mine
11. Rowley mine
12. Copper Butte
13. Green Mountain mine

• Copper occurrences

▲ Assays only (with 1% or more copper)

NOTES

Pictured is AL the owner of the mine

White Horse Creek Mine

N.42° 47'52.8"

W.123° 09'68.8"

Elevation: 1689 FT.

The White Horse Creek Mine

This one is a must see. I have the permission from the owner for all to look and try to find it. Please leave it better than it was. If this one is destroyed the owner will lock it up for good. Please take care of it!

There is a lot to see in this one! Quartz, Iron, Shell, old half eaten bones, and, oh, Bats too. That's right this one has very large bats. If you look at the picture below at the ceiling of the mine, that black spot is a bat just getting ready to fly at me.

The mine is more than 200 feet deep and has three short tunnels. The first tunnel is 75-feet in, then the mine "Y's" into two shafts about 150 feet. One tunnel is about 25 feet long and the other about 75 feet deep.

Be sure to bring a powerful flashlight, and if you can get one have a red light available. With the red light the colors will burst into a symphony of lights. In addition, the bats look real neat in red light. Here is another couple of pictures of this one. It's too nice for just one photo.

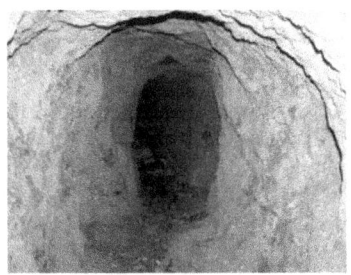 

Notice in the picture on the left the black spot on the roof, that's a bat. The picture on the right is taken from about 80 feet inside the tunnel looking out to the entrance, before going around a bend in the shaft, and losing site of the opening. Boy did it get dark quick. There is so much to tell about this mine, and no room to tell it all.

According to the owner of the mine now, the original record shows a man named "X" opened it in 1878. This is all the information I have found on the name. However, we do know that more than 1.2 million dollars worth of gold was taken from the White Horse Creek mine before Mr. X died in 1904.

To this day no one knows where he hid all the money he made. Every month he took the gold to Canyonville or Glendale to sell it. He would spend a few days buying supplies and leave. He didn't spend much of his money, before leaving town for good.

The new owner started asking questions about mining. He was told the story about the old man and the White Horse Creek Mines. The new owner searched the Whitehorse creek area for some 9 years before finding them. He and a partner started working the shaft in 1972. They have added another 60 feet onto the length of the shaft but, have yet to find more Gold.

Enter at your own risk. And again, please leave it as you found it! Please Please Please be sure and replace the spring mattress in front of the opening! This will insure the next visitor will not walk into a bear or cougar.

Driving Directions:

Take exit 188 off I-5 and go towards Galesville Dam, go about 4-1/2 miles to Longfibre Park, turn right, follow Longfibre about 2 ½ miles to a campground on your right. Stop here and walk. If you park in the wide spot in front of the camp ground, go across the road and climb up the side of the hill about 100 feet. It is a very steep climb but well worth it.

We're not without Pirates

The Douglas, Coos, and Curry counties are not without their own tales of Pirate ships.

The one that has probably been talked about the most is an unknown pirate off the coast of Oregon during the mid to late 1800's.

It is said that he would attack the ships coming from San Francisco on they're way to Seattle and Alaska.

The tale says the pirate ship would attack near dusk. Then it would run for a cove near Coos Bay.

He would take his spoils from the attack and hide them in a cave.

According to legend the cave is in a valley 2 miles from the cove. Hidden about 500 feet above the ground in a wall cave.

The story says he could watch the patrol ships go by and see his ship from the cave.

If a patrol ship was sited, he was able to return to his ship and leave the cove for the open seas before the other ships could arrive.

The legend tells of all the gems, diamonds, rubies, gold, and silver he collected over the some 20 years in operation.

The ships demise came when the local coast guards started a rumor about a huge shipment of gold, silver, and cash from San Francisco was on its way to Seattle. Carrying on board the railroad, miners and loggers payrolls. It didn't take long for the captain of the pirate ship to hear of this.

The local coast guard didn't let out that there would be two ships, one decoy, another full of men and guns to stop the pirate.

The faithful day came and the pirate ship was waiting behind a large rock just offshore. When the first ship went by he raised his sails and set out after it.

A battle ensued, and as the men aboard the pirate ship crossed over for they're loot the second ship and all its manpower came up without notice!

They fought a useless battle, as both ships were sunk to the bottom of the ocean. The captain went down fighting and never gave up his name or the place he hid his folly.

To this day every time I drive down 101 from Coos Bay to Brookings, Oregon I think of the tale of the Pirate of the West.

I also watch the hills and valleys for a possible glimpse of a sign for the cove and the cave. The kids love the idea of looking for treasure while on a road trip, also!

This story has been told repeatedly. But, unlike other yarns this tale never changes.

But, no one knows *where* or *if* the Cove and Cave exist!

The Legend Of
Two Bucks and Little Doe

Two Bucks and Little Doe were Cow Creek Indians. They were not related, but they were born on the same day. Little Doe was the first daughter of one of the Chiefs of the tribe.

Two Bucks family, while known by the tribe, was of no affluent background.

Little Doe got her name because a doe deer was seen just outside the opening of the tent when she was born. Two Bucks was named this because his father was out hunting when he was born and came home with two (actual) bucks.

Two Bucks and Little Doe grew up together, inseparable almost. Even through the "I don't like girls ages" they were always friends. As they got older they knew they would be married someday if Two Bucks could come up with the dowry. Little Doe being the Chiefs daughter the price for her hand was large. If Two Bucks could come up with the price needed for Little Doe's hand he would become the chief when her father died.

The legend is said that Two Bucks knew the white man paid a lot of money for gold and silver. He could find enough, sell it to the white man at the trading post, and buy the horses and cattle he needed for Little Doe's hand.

It is said that he worked and saved for 12 years. In this time he had saved more than 25 pounds of gold, and 10 pounds of silver.

He thought that by his 24th birthday he would have enough to sell to the trading post and get the hand of Little Doe.

Little Doe's father knew this and had thought hard about Two Bucks becoming the chief someday. When Two Bucks was just a few months from his 24th birthday, Little Doe's father agreed to let another brave pay the price for her hand.

When Two Bucks and Little Doe heard about the arrangement they ran away together. The chief sent the tribe looking for them to bring them back.

When they arrived back at the camp, Two Bucks took out a knife and killed the Chief. He then killed himself.

Little Doe, seeing what had happened and not wanting to spend her life without Two Bucks, took the knife and ended her own life.

It is thought that Two Bucks had hidden his gold and silver in the valley of the "Indian Shower". This is about two miles from a Camp in Drew, Ore. The Indian Shower is two miles up Tommy creek just off Tiller Trail Hwy at Drew.

This is a story that was told to me by Running Elk. Running Elk is a 87 year old Cow Creek Indian who remembers the story of his Uncle "Two Bucks" and his love "Little Doe".

The No Name Mine

N 42° 54' 80.7"

W 123° 08' 82.9"

Elevation: 1645 FT.

The No Name Mine

There is not much I can find out about his one other than the owner on file never named it and signed the claim with and "X".

It was opened in 1897. More than $25,000.00 worth of gold was taken out. No one knows what happened to the owner. He just packed up and left one day.

No body has been inside for many years now. It took me more than four hours to even find it! The over growth keeps it covered.

There are some very large trees and bushes in front of the entrance as you can see by the picture. I tried to clear a better opening, but the brush was too thick.

A creek runs about 100 feet in front of the entrance. When I stopped to do a little prospecting, I found the creek is loaded with color. The gold was small to medium size flakes and were very rough to the touch. This means the lode is not far up the hill!

Cherokee Mine

Opened 1905

N 42° 52' 10.3"

W 123° 10' 53.3"

Elevation: 3364

The Cherokee Mine

This mine makes you feel like you're on top of the world. The view is fantastic! If you park your car in the turnaround area you will see what I mean.

The Shaft is between the upper and lower roads. **Be very careful as the shaft goes straight down about 100 feet!**

The tunnel is down the hill to the south from the shaft. When you look down the hill you will see a five foot wide trench about 60 feet long.

This was used to bring the ore out of the tunnel. The tunnel opening is about 80% caved in, but I am able crawl into it. The tunnel runs for about 100 feet and makes it about 50 feet from the main shaft.

The owner gave up at that point and left the mine. This one is a great place to spend some time and have some fun. The view, wildlife, and family fun will be well worth the time spent getting there.

Driving Directions:

From I-5 take Exit 88 towards Galesville Dam. Follow the main road to the dam. The lake is on your right and you should see the boat landing and service road. There should also be a gravel road to your left. Turn left and stay straight. You will get to a junction about 3 miles up. There are four roads that come together. You will turn left on the road that switches back along side the road you just came up. Follow this one about another three miles to another "three way" junction. Take the left fork. Follow this road for another one mile you will level out and stop.

The Lost Camas Valley Gold Mine

This is a true story! Not a tale or legend. Harry lived for most of his life in a small town on Hwy 42 west of Winston Oregon called Camas Valley.

A quiet little town where few people worked in town. They all had to drive to Roseburg or Winston. Harry was a fixture of the town and everyone knew him. They would always see him walking around the streets and valleys in the area.

Little did they know that when Harry left his home he was heading for his secret gold mine. The mine was close enough to walk to, but he never took the same route. So, no one ever knew where he was heading.

He was able to leave home in the morning and be back home before dark with enough gold to last him six months!

It seems for years he had many so called friends trying to find out where the mine was. But, he would never tell. Now, the rest of the story is about my own family member who was a true friend of Harry's. He never asked to help work the mine or even ask where the mine was, he didn't want that to come between them.

Harry had a friend that owed him money and would never pay. He would ask and ask but to no avail. One day while this friend was at the house he and Harry got into an argument.

They had both been drinking and had really tied one on. Harry had gotten so mad he went in the house and got his pistol. When he came back outside his friend was still trying to fight. He came at Harry with a pitchfork so Harry shot and killed him.

Harry took the body somewhere it would be found but not for a while. He then hid his gun. A few weeks later the body was found.

Everyone in town knew Harry and this guy had been feuding for some time so obviously the police came to question Harry. He told them he knew nothing.

The Sheriff on the case didn't believe Harry. He knew how bad his temper got when he had been drinking. The investigation went on for years.

The Sheriff searched Harry's home, land, cars, trucks, everywhere. But, the gun was never found. The whole time Harry was laughing at them.

Ten years had gone by and the Sheriff gave up his search. Harry was getting old, and was in bad health. His one true friend still never asked about the mine and still didn't want to know about it.

Harry told my family member that he wanted to tell him where it was. But, he told Harry he and didn't want to know.

About two years later, Harry was in the hospital and things were not looking good. The doctor told my family member, that Harry wouldn't live through the night. Harry called for his friend to come in and visit. Harry had every intension of telling his friend where the mine was.

He did tell him that there is enough gold there for him to live a very comfortable life.

Harry was told again, "I am not your friend because of the gold". He told Harry to keep it to himself. He did however give a few hints to where it is. He told his friend that it was a 1 ½ hour walk north east from his home, then follow the ridge to the split in the mountain. Look North you will see a giant boulder. Harry died. His friend never even went to look for the clues.

To this day the mine is still not found and is on BLM property. Good Luck!

The Gun Fight at Bear Creek Mine

This is a sad story about the gunfight at Bear Creek near Camas Valley. Bear Creek is at the bottom of twelve-mile hill on hwy 42. A memorial has been etched in stone for the gunfighters.

William Holcomb and his son William "Billy" Jr. were working a mine in the bear creek area. One day they went to town for supplies. When they got to town, they heard talk of a man by the name of Guy Mallard had been claim jumping all over the area. They soon found out that this guy was trying to put a claim on their mine also.

William Sr. came across Guy and confronted him. Guy told him he had every right to file claim on his claim. He told William to leave him alone and move away from the mine.

William told him he would die before letting him take what is ours. The man said that could be arranged. Fearing for his sons life not to mention his own, he and his son packed the horses with supplies and went back to the mine.

The fighting and feuding went on for more than a year. Then one day the stranger came to the mine to confront the father and tell him once and for all to get off the property that was now legally his.

The stranger stood back and challenged William to a gunfight. William Sr. was not a gun fighter and didn't think this guy was one either. Billy was working in the mine when he heard the gunshots. He came running down the hill to see what was happening.

When he looked from a crop of rocks he could see his father lying on the ground shot dead by Guy. Billy, 21 years old, 6 feet tall and lean with light brown hair like his mother. He knew how to shot a gun. Billy waited in the rocks until Guy left to go to town. Why he didn't look for Billy, no one knows. Possibly he needed him alive to tell where the mine was.

After Guy left, Billy went down to see and bury his father. He then went to town to spread the word about what had happened. The town was outraged and everyone started looking for Guy. William had helped everyone in town at one time or another and was liked by everyone. When Guy heard about the rumors of him killing William in cold blood, he went crazy and swore he would kill Billy too.

Billy went back to the mine to wait for Guy. When Guy got to the mine his gun was out and ready to shoot. Billy figuring Guy would try to shoot him in the back had taken some clothes and made a decoy. Guy opened fire on the decoy shooting him 6 times in the back.

Billy stepped out from behind some rocks, just behind Guy. Billy said you missed. But I won't! Guy turned around trying to get at least one shell in the chamber. Billy pulled his gun out and pointed it at Guy. Guy started begging for his life. Billy said, "You don't deserve to live" and started shooting. A few shots later Guy laid on the ground dead.

Billy buried Guy next to his father. Now the story turns, a few years later. Someone buried Billy next to Guy and his father, and scratched on a stone who killed who. Who did it no one knows. Where's the mine? No one knows that either. But William and Billy had found allot of gold when they worked it, so its still there today. This is a very true story so if your ever up Bear Creek outside of Camas Valley check it out.

The Chinese Gold Vein

In 1878 while the Chinese were building the railroad line from Roseburg to Grants Pass Oregon, when they built this tunnel they found a large vein of gold right in the middle. The Chinese workers would work all day on the railroad then spend a few hours at night digging in the side of the tunnel. The railroad found out and wanted to stop them so with all they're wisdom they have the Chinese workers seal the tunnels walls with concrete.

The Chinese being resourceful they just dug a tunnel behind the Railroad tunnel. Today this area is known as Peck. Coming from what the Railroad called what the Chinese were doing they were Pecking at the walls of the tunnel. You will see this Railroad tunnel on your way out of Riddle, Oregon. It is about 12 miles up Cow Creek past the Doe Creek Bridge.

The Quines Creek Mine

N 42° 43' 09.4"

W 123° 14' 06.1"

Elevation: 2848 Ft.

The Quines Creek Mine

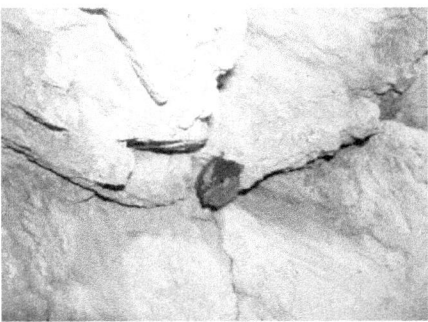

It's hard to think of what to say about this one. I have seen more equipment here than at any other mine or quarry. The Tunnel is a hard fine but well worth the look see.

Bat's were everywhere in this one, most were very small. This mine was a great producer of gold, so big in-fact that the owner hired a guard with side arms to watch the place full time.

Directions

To get here take exit 86, off I-5 Quines Creek, turn east just past the restaurant. Follow it past Bull Run Rd. about 2 miles. The road will fork stay left, do not follow the signs for Quines Creek Rd. as this will take you into private property. At the next junction stay right, about another mile you will come to a crossroad.

IF you turn right you will go back into the private ranch, if you go straight you will end up over the hill into Wolf Creek. You will want to turn left, as soon as you turn there is a 4-wheel drive road on your right. Take this one to the first landing, you will see all kinds of equipment. From here I suggest walking the next four levels.

The Hole in the Wall Mine

This is the Rouge River Hole in the Wall. Out of Wolf Creek to Galice you will cross over the Rouge River Bridge. Just about a half mile from the bridge is the hole in the wall. Years ago the county came along to put the new road through as you see it today, problem was they had to remove more than 90 feet of the mine shaft. The mine now is only about 15 feet deep.

This is a great place to spend some time. A wonderful hiking trail is on the west side of the Rogue River. There is a great recreational area on the East side of the river. Restrooms, Raft launch, beach area. Just 100 feet away is the mouth of Graves Creek!

Graves Creek is a great place to pan and sluice. You can only use the first 100 feet of Graves Creek due to all the open and active claims.

NOTES

The First Strike Mine

N 42° 52' 81.2"

W 123° 21' 33.9"

Elevation 2603 FT.

The First Strike Mine

I love this one. Once you get past the first 100 ft. it opens up into a very nice solid tunnel. *The first section is very dangerous be extremely careful!*

Opened in the late 1870's and producing more than $3,000,000.00 in gold on record. The mine closed to the astonishment of everyone in the area. The mine was still producing more than its share of gold and gems.

The owners of the mine, due to health, decided to stop operations in 1919. Since then there has been more than twenty owners of the claim, all have produced a good share of gold. The tailings piles produced a abundance of gold flakes for me when I took the time to crush some up.

The Legend of the Wolf Creek Gold Cache

This story is stranger than true. However, the story is true to everyone who knows of it. It all started in June 1863 when a poor miners son set out to find his own treasure and prove his father wrong. His father had told him for years that he was no good at being a miner and should go to school to get an education and find a good job.

Hugo Rancher was born and raised in Prospect, Ca. He had his mind set on being a miner and find the mother lode. Therefore, when Hugo was 15 years old he packed up and left for the Northwest. He traveled north and staked a few claims along the way, but, he never found much. Antsy to strike it rich he would move around a lot.

One day in 1865, Hugo came to this little mining town. He stopped for the night at the local motel. After getting a room and taking a bath, he dressed and readied for dinner. He went down stairs. He had no idea what was in store for him this evening.

He sat himself at a table. The waitress came over to take his order. This was it. All he had to do was hear her voice, and he knew she was the one for him. He looked up and saw this young girl of 16 starring back at him. Hugo asked her name, s "Mary Ann", she replied. As she looked at him, a new sparkle shined in her eyes. It was love at first site.

This is where it all started. Hugo wanted only to make Mary Ann happy. He told her he was a miner and had to leave for a new prospect area. Hugo told her he would be back in the fall and asked would she wait for him? Yes she replied eagerly.

Hugo set out the next morning with a new reason to make it big. He had forgotten about the things his dad had told him all those years. Now it was for Mary he was going to make it rich.

The first year went by and when Hugo returned to Wolf Creek, he found Mary waiting for him as she said she would. He told her he had found their pot of gold and were going to be very rich. He told her of a place outside of town at the switchback of Wolf Creek there was a tree that had been split by lightning, fifty steps north of the tree was a bolder the size of a horse. Behind the stone was a small opening that went under the bolder about three feet.

Hugo went on to tell Mary that there was a bag full of gold nuggets, telling her it weighed about 50 lbs. He then pulled a small bag out of his coat pocket and handed it to her, it was full of gold. Well worth more than a years salary for Mary. He then asked her to marry him. She without hesitation said yes.

They were wed the very next day. They spent more than a month together when Hugo said I must get back to the mine. Mary didn't want him to go, but she understood it was his job.

Hugo set out for the mine the next day. He told Mary he would be back in a month or so. This time he didn't keep his promise, while on the way to the mine two men jumped him. They beat him for days trying to get the location of the mine and the cache of gold he had hidden. Hugo soon died from internal bleeding. The two men went back to town and did the same to Mary Ann, she never told them where the mine or the bag was hidden. Therefore, to this day somewhere up Wolf Creek at the switchback is a split tree near a huge boulder. You know what you have to do now!!!

The Caches of
Gold at Days Creek

The legend is told of thirteen caches of gold buried at or near Days Creek, Oregon. The year is 1878 the fall was near and the miners were in town for the weekend. While at the local salon, they were told the tax assessor would be up to their claims by the end of the month.

When the miners went back to they're cabins, they had a meeting and decided to hide their caches only keeping out enough to show they had found some gold during the summer season.

They each decided to bury their gold near their cabins. Each miner picked a place when no one was looking. They paced off 25 paces from their front doors and buried the caches near a marker on their claim.

A few weeks went by and the assessor hadn't showed. The winter was really starting to hit hard. The men worrying about the taxman showing up they waited and waited for him. Then over night the weather changed and caught them all off guard. It snowed more than three feet over night.

When the miners woke up the next morning, they could hardly see out the windows of their small cabins. They new that the tax assessor would never travel in this kind of weather. However, the ground was too cold to try to dig they're caches up. They all decided to wait until spring to dig they're gold up.

It turned out to be one of the worse winters in history, more snow than the area had seen in many years. Three of the miners died that winter from freezing cold and the flu. Two more died from pneumonia, but the all told their closest friends where their caches were buried.

When spring came it warmed up so fast Days Creek looked more like a raging river. The waters rose and flooded the area. Taking trees and a few of the cabins that was too close to the creek. The miners that survived the winter barley made it to higher ground before the floods.

When the floods stopped and the miners were able to get back to they're claims they found not one cabin still standing. There were no markers as to where their cabins were the creek had changed course during the flood.

The miners spent the next several years trying to find they're caches, but no one ever did. To this day the old timers of Days Creek still talk about all the gold that was buried along the creek. In addition, to this day no one has ever found it either.

Most of the area where the miner's cabins were is now private property, or is it? According to old maps the creek's course before the flood would put it not on private property but in BLM property.

So if you ever get up in this area be sure, look at an old map of Days Creek, and see where the creek once ran. You may be the one that happens on at least one cache of gold.

Doe Creek Mine

N 42° 55' 73.8"

W 123° 30' 39.0"

Elevation: 1265 FT.

The Doe Creek Mine

This Chromium mine was opened during the 1940's.

Along with the Chromium there was more than $10,000.00 worth of rubies taken out.

During its short life there was a very large pocket of Chromium, but it ran out soon. The tunnel is only about 40 feet deep but the colors of the many different ores are still very visible.

I would like to tell you more about this one; I can tell you that just 30 feet away is another shaft that tunnels in more than 200 feet. The tunnel opening is caved in and no one has been able to find the other shaft.

Directions:

I-5 exit 103, go west, past all the lumber mills. About 7 miles you will come to a one-lane bridge, turn right just before the bridge. Go less than ½ mile, and the tunnel is on your left, just before starting up a small hill.

The Placer Mine Cabin

The Placer Mine Cabin

The original cabin was built on this spot in 1879.The new cabin pictured here was built about 1923. The new owner of the Placer claim wanted a newer bigger cabin to live in. According to all information I can find about the Place Claim, it produced about 40 plus pounds of gold between 1880 and 1921. After the new owner took over there has not been much found or at least talked about.

The Cave-In Mine

This mine is closed to the public because of the danger of more landslides. The whole side of the mountain came down some 40 years ago. Since then the slope has washed out many times. As you can see by the picture above the tunnel is not very safe. I would love to tell you where it is but I can't give you the exact location.

BUT, If you go to the Iron Door Mine all you have to do is go south west about a half mile, the road Y's now I am not going to tell you that if you take the road down hill at the "Y" you might just come across The "Placer Cabin" and just past that you might find "The Cave-In Mine".

IF you get to this one, watch your step when you enter the cabin. About four feet from the door is a shaft leading straight down about twenty feet! **So be careful!** it's a long trip back to town to get help.

The owner of the "Placer Cabin" mine, wanted to mine in the winter out of the weather.

The terrain is quit rough here. But, there is a lot to see.

At the time this book was being written the claim was available. It is so hard to get to though nobody wants the trouble. The picture below right is of the shaft (straight down!) in the middle of the floor inside the cabin. **So watch your step!**

The Tail of the Two Brothers Mine

This is a story that was told to me by an old gentleman from Galice Oregon.

He said the two brothers had found their pot of gold not in the mountains of the Rogue River valley unlike so many others. They had found a vein of gold right in town! A place no one had ever thought of looking.

It is said that underneath one of the old buildings in downtown Galice is a shaft that leads down to the tunnel of the Brothers Mine. The shaft is three feet round and 48 feet deep.

Once you lead you in more than 300 feet. Along the way there is more than 12 adits of 200 feet and more. However, you need to find the right adit to get to the mother lode. More than 15 people tried to find the main adit to get to the gold and they were never heard from again!

If you can stay in the main tunnel by taking the right adit, it will bring you to a large cavern of Quartz and Greenstone. Among the Greenstone and Quartz you can see the veins of Gold. It is supposed to still have all the Gold you could ever want.

Can you find the right building and get permission to crawl under it?   **CHALLENGE!**

A Mine of his own

This is the only thing little Jerry Holcomb could think about. He saw the rich men in town walking with their fancy cloths and talking of the amount of gold their mines had produced. Jerry could think of nothing else but getting a mine for himself. When Jerry turned 14 he did just that. He spent the next eight years prospecting the Rogue River, Galice, and Wolf Creek areas.

One day when Jerry was almost turning 23 years old, he showed up in Wolf Creek. He was carrying a small but very heavy carpetbag. He went directly to the assayers office. Jerry had done what he set out to do. He had found his own mine and struck it rich!

Jerry Holcomb had found an old Chinese mine that had been left for better diggings. What the original owner didn't know they had dug the tunnel just off the vein of gold. Jerry worked the mine for more than three years, and had made himself a very wealthy man.

Now at the age of 26 he had enough money he could find a wife and live very well for the rest of his life. He had his eye on a young lady in Galice, she had red hair, green eyes, and the most beautiful face he had ever saw.

She didn't know he had struck it rich, until after they were married. Once she found out she wanted him to go back to the mine and get more gold. She wanted to be richer than she was. Jerry wouldn't go or tell her much about where the mine was, just that it was just a mile west of Wolf Creek one-quarter mile south of Graves Creek. He told her nothing more.

She was so upset at him she had some friends beat him until he talked. However, friends of Jerry's did not let her nor her friends find the mine. To this day it has never been found.

Beware! Gold fever is a dangerous mistress.

The Lost Gold of Galice

This tale dates back to long before the gold rush had come to Oregon. A tale of a miner named Louie Pratt who had been prospecting all over the western United States.

In his 78 years of life he had been in places that no other man had ever traveled. In 1835 his travels brought him to what is now Galice, Oregon. Although at the time was just wilderness. Full of Wildlife! FULL OF DANGERS!

He found in the high mountains just outside of Galice a large over cropping of stone about 500 feet up the side of a mountain. The hill faced south towards Galice, and had the evening sunlight shine through a small hole in a hill to the west.

Just like the legend of the "lost Dutchman Mine" in Arizona. The beam of light pointed to a place that was solid gold. The legend tells it that the gold was an area of more than 3 feet wide and 60 feet long. No one knows how thick it was.

To this day if your driving out of Galice towards Wolf Creek, at the right time of day watch the mountain side for the tracks that Louie Pratt had made. It is said that he had worked the area for more than two years when he died on the trail back to his cabin on the Rogue River below the ledge.

His gold and the ledge have never been found. But, you can see some flickers of sunlight on the mountainside when the sun is setting. It seems to be more obvious in the late summer months of September and early October.

The Winter Treasure of Hold Up Canyon

It was in the middle of fall 1873 when Jeremiah Wilkerson decided to head for the mountains above Cave Junction Oregon. His friends and family tried to get him to wait till spring but he wouldn't have it. He told them if he got started now he could be set before winter settled in.

He started out heading northwest before daybreak in the middle of October. It was a very cool morning and the clouds were thick and heavy. No rain yet but most could tell it wasn't far off.

The first day Jeremiah had walked and packed his supplies almost 25 miles. It was very rough terrain scattered with ditches, creeks, and signs of mountain lions and bear. He had to really watch the surroundings for signs of trouble.

Indians were not much a problem by this time but a few renegade tribes were known to be in the areas he had to travel through. By the sixth day, exhausted from the pack and terrain Jeremiah decided to stop early and get some rest for the rest of his trip. His plan to work his way northwest another three days to the valley of Bear Canyon was starting to feel like a dream not a reality.

After setting up a temporary camp, Jeremiah built a fire so he could have his first hot meal in a week. He shot a rabbit earlier that day for food. After eating his fill he hung the remaining food up high in a tree so not to attract any unwanted animals.

He then settled down for the night for some very needed sleep.

It had turned quite cold that afternoon, and the smell of rain was getting stronger. During the night the rain began. It came down like a river!

Jeremiah's temporary camp had a torrent of water flowing right through his bed. Somehow he was still able to get a few hours of rest.

The next morning Jeremiah found himself on the bank of a small un-named creek.

He started looking around for a place to set up a camp for a few months knowing that the rains would have made traveling any further to dangerous.

He found a small cave it was 30 feet deep and more than big enough for him to make a very nice covered campsite. Hard rock above made this a quite safe place to hold up for the winter.

Jeremiah started moving all his supplies inside the cave. He then built a door to cover the opening to keep animals out.

Ventilation for a campfire was going to be no problem for the way the cave was built the smoke would rise along the edge of the cave to another opening above him. He set out to make it as comfortable as possible. Jeremiah figured he would be here for about five months.

After a few days of just sitting and listening to the rain, Jeremiah was getting very bored. He started looking over the cave and some of the small tunnels leading off deeper into the mountain. There were five small tunnels that needed to be explored.

One of the tunnels led Jeremiah some 200 feet before ending at a rock face of quartz. After spending more than a week digging he found nothing of real value. He searched out two more tunnels and still found nothing.

Almost two months had passed sense Jeremiah left his home and family.

The first snowfall had started so Jeremiah spent a day gathering wood for fires and setting traps for small game. He was throwing more wood on the fire to get the cave warmer when he heard something outside.

Jeremiah grabbed his gun and headed towards the noise. He saw laying on the ground a few feet from the cave a man who looked like he was hurt.

Cold from the weather and unable to walk Jeremiah picked him up and carried him into the cave.

Once Jeremiah could see the old man was an Indian. Old, hurt, and cold.

Jeremiah covered him with a blanket and took care of his wounds. After a few days the old man started to come around. The Indian was obviously feeling better and able to sit up on his own.

Jeremiah took care of the old man for more than two weeks. Neither of them knew how to speak each other's language. Their they're names were the only words they could understand. "Walking Eye" the old man was able to say. Jeremiah wasn't that lucky. The best the old man could do was Jerma. But, this was more than enough for Jeremiah.

He hadn't realized how much he missed people and having someone to talk to. He would sit all day talking to Walking Eye even though he knew the Indian didn't understand a word he was saying.

After a few days Jeremiah went off into one of the tunnels. After digging around for a few hours he came back with a smile on his face. He had found some gold in the one tunnel.

When he showed Walking Eye and started talking about looking for gold, Walking Eye interrupted him. Walking Eye started talking and talking, with some excitement and a smile. Jeremiah wished he could understand him. Walking Eye took Jeremiah's hand and led him out of the cave to another cave about 100 yards away.

Walking Eye got down and crawled into this little opening, Jeremiah followed not knowing where they were going but following close behind. After a few feet the cave opened up into a large cavern. There were tunnels leading off in every direction!

Walking Eye led Jeremiah down one of the tunnels and showed him a vein of gold that was more than one foot across and as long as the tunnel! Walking Eye held out two hands and pointed to Jeremiah as if to say you can have it.

Jeremiah couldn't believe how much gold he was seeing. He looked at Walking Eye and bowed his head to thank him. With a smile on his face Jeremiah could see how happy Walking Eye was to see that the gift was well accepted.

They soon went back to the camp for something to eat. They spent the rest of the day just rambling not knowing what the other was saying.

That evening they lay down for the night. Jeremiah could hardly get to sleep thinking about his find. At daybreak Jeremiah got up to fix breakfast and noticed that Walking Eye was packed up and gone.

Jeremiah looked all around for him thinking maybe he had just gone out for a walk or look for something to eat. Walking Eye was nowhere to be found.

When Jeremiah got done with his morning chores, he headed out for the other cave. He started looking around and there were no signs that anyone had been inside the other cave. There were no markings or disturbed ground showing that Walking Eye or Jeremiah had been in the other cave.

When Jeremiah crawled in he went straight to the tunnel that Walking Eye had showed him. The gold was everywhere! The gold was as thick as the heel of his boot. A vein as long as a 10-mule team, little did Jeremiah know the vein was as wide as a stagecoach.

When he started digging in the wall of the tunnel he couldn't believe his eyes. The gold was so thick it went in as far as he could reach with his arm and more.

Jeremiah spent the rest of the winter digging in the tunnel. When spring came he loaded up a bag of gold to head to town and stake claim to the land. He had found much more than he could carry so he buried the cache out side the camp cave. Then he covered the opening of the gold cave so no one would think to look for it.

When Jeremiah reached town with his gold all his friends were there to see him and see how he did. He obviously kept the location of the Gold real quiet. He only told those who he knew he could trust.

Word spread fast in the little town of Cave Junction. Within a few days word had spread to every nearby mining town about his find.

Jeremiah loaded up on supplies to head out to the mine, he left heading in a northeasterly direction to throw off those who might try to follow. But, follow they did. When Jeremiah reached the top of a hill and looked behind him he could see a group of miners following him. He walked for more than four days until he came to an old cave that he knew well.

He scuffed his feet to make it look like this was the area he had been digging. He went in and quickly built a fire. He left some of his supplies scattered around and then he went above the cave and waited.

Soon the group showed up.

He watched them enter the cave with their Guns drawn.

When they were just inside Jeremiah pushed a big pile of rocks down causing a landslide. It covered the entire entrance to the cave.

Jeremiah knew this would only slow the miners down. But, it would give him time to double back and head to the real mine.

Jeremiah spent the next year digging and hiding most of the gold he found. Glass quart jars filled to the top with gold was hidden 50 feet from the cave entrance of the camp cave.

Trying to leave no traces as to where he was hiding the gold Jeremiah covered the ground with crushed quartz rock. By doing this he was able to cover his tracks because the quartz rock is white.

Soon Jeremiah had enough gold to head for town. He packed up his mule with more than 20 pounds of gold leaving the rest of his gold hidden at the mine.

It was a warm Sunday morning in June when Jeremiah left the mine for town. With the mule loaded down it was going to take him an extra two days to make the trip. Jeremiah figured he wouldn't reach home for a week or more. He didn't want to over work the mule.

Jeremiah had a lot on his mind as he traveled. How this gold would help his family out, etc. . But mostly about the girl he left behind.

He prayed she was still there.

On the third day of his trip he noticed the mule was not doing well. He first thought he had over packed and the mule and it was just tired. But, by the end of the fifth day the mule was barely able to walk.

Jeremiah unloaded the mule early that day to see if that might help. He repacked most of the supplies into his own pack. On the seventh day the mule laid down and died.

When this happened Jeremiah loaded all the gold and as much supplies he could carry.

He got an early start the next day.

Travel was very slow due to all the extra weight. He knew he had to have the supplies to just survive and wanted to have the gold for his family. Now more than 6 days to home at the rate he was able to travel he saw one of his Indian friends he had met through Walking Eye.

The young Indian saw the trouble his new friend was in and offered to help lighten the load. Jeremiah gladly accepted his help. They made good time for the next two days, but on the third morning when Jeremiah woke up at sunup. The young friend was gone. The young Indian has left Jeremiah with only his gold. Almost all his supplies were gone. Not knowing what he was going to do with at least the supplies to survive he just started walking. Breakfast, lunch and dinner came and went.

For two more days it went like this. Jeremiah had not seen any game running around. It was like something had spooked them all.

He was getting very tired and sore from carrying all the supplies. Jeremiah was now four days to home. He figured how many supplies could he drop and still make the trip. He started dropping supplies to lighten the load. It helped the first day. But, his sore back, arms and other trail related ailments were really getting to him.

100

Jeremiah was now stopping to rest every hour. With each stop he made he would drop more supplies.

His pack was now down to about 150 pounds but, his back ached with pain and his arms felt like they could just drop off. He was getting tired of the stress and strain his body was under.

After one more slow day and stopping early so he could try to relax and get a good meal. Jeremiah still could not muster the energy to load his pack and start another day on the trail.

Jeremiah realized he was only two days from home! This seemed to give him a little more strength. His stride picked up and started thinking about how the Gold will help his family and all he was going to do with it.

The first thing he thought was get a horse. Then he would buy a years worth of supplies for the mine. He also thought about hiring a helper to do all the odd jobs around the mine, including building a new cabin.

The day went very fast, when Jeremiah decided to bed down for the night it was almost dark. He would have to hurry to get a camp set up.

He went to sleep with out anything to eat, just thinking about being home tomorrow afternoon. When Jeremiah woke the next morning he had quite a hunger, he built a fire to fix some coffee and breakfast. After he ate he loaded his pack thinking this is the last time he would have to do this.

He started his trip after about three hours he could smell bread baking from town. When he reached a small hill outside of town he looked down and saw his home.

Jeremiah almost started to run. His step was like the first day he set out for home almost two weeks earlier. He stopped at the assayers office to get a total on his gold.

Jeremiah told Bill the assayer to just take the total to the bank for his account. Bill was very happy to see him and started to tell him about the miners that had chased him. When they arrived back in town they told everyone you had been killed in a landslide. They never found your body but looked for it for several days.

Jeremiah couldn't believe what he was hearing. Why did they chase me he asked Bill? Bill told him they told everyone they wanted to help you with your claim and work for you. Jeremiah said is that why when they saw me go into a cave they followed with their guns out and shooting when they went in the cave. Jeremiah told him they were out to kill me not help. They wanted the mine for their own.
Bill told Jeremiah who the men were and asked him to not tell anyone he was back in town until that evening. Jeremiah knew that they would all be at the saloon after dark.

Bill agreed and told Jeremiah how much money he had earned with the gold he had brought back. Jeremiah told him to take the totals to the bank and go straight to Mr. Johnson the manager of the bank.

Bill agreed.

Jeremiah laid his pack down in the back office. He pulled his hat down low so no one would see who he was and headed for his house.

When he got to the door he knocked and his mother and father came to see whom it was. When they saw that it was Jeremiah they couldn't believe it. Jeremiah told them about what really happened with the miners.

That evening Jeremiah went to town and straight to the bar. He saw all the men there that Bill had told him about. He entered very slowly and quietly. When Jeremiah was standing right behind all of them, he pulled his hat off and said "hello boys".

When the men turned around and recognized who was speaking, they became noticeably nervous. The Sheriff was standing right there with Jeremiah.

The Sheriff asked the men "why did you tell everyone Jeremiah was dead. You boys didn't follow him to help him you followed him to kill him".

One of the men started for his gun but the Sheriff drew his gun first and stopped the man.

The Sheriff told the men to drop their guns. He told them they were under arrest for attempted murder. The men dropped their guns and walk to the jail.

Jeremiah yelled out "I'm home! Drinks are on me"!

Everyone was, needless to say, very happy to see Jeremiah. Why is this story is being told?

It is said that Jeremiah, while he was home, married his love Amanda. He never got time to go back to the mine. His treasure still lays buried in front of the Winter Cave!

ADDITION:

Now the story is, Jeremiah traveled northwest out of Cave Junction 9 days going toward Bear Canyon.

He never made it to Bear Canyon but was three days from it.

This would put the Winter Cave Mine in the area of Lawson Creek in the Quosatana Butte Range.

The area is perfect for the gold and caves that are talked about in the story. I have searched the area many times. There are so many caves in this area that trying to explorer each one would take years.

I would love to have any more information on this area and if some one might know of the story and more details!

Good lock searching this one. You'll need more than just a few days!

The End of My Life Cache

This is a short story that was conveyed to me more than twenty years ago. I was talking with a friend an old miner name Rascal Pete.

He told me many stories of lost caches in the Jackson County area. This is one that will set your heels on fire to go look for. It dates back to 1881 when a band of Indians were trading with the locals around Jacksonville Oregon.

The trade beads and pelts were not enough anymore to get the trade they wanted. The local stores were overloaded with such items. They were not selling as fast as they use to.

The locals told the Indians that they needed Gold for trade! Not just beads and pelts. So the Indians started bringing in gold and a lot of it. They did not know how much it was worth at this time. And of course no one was obliged to tell them.

The locals took in pounds of gold for horses that a few ounces would have bought. The Indians would bring in a pound of gold for a bottle of whiskey. Another pound would buy them a new rifle!

You can see that the Indians knew where to get the gold but kept it secrete so the white man couldn't find it.

A trapper set out to track one of the Indians to a cave not far from Jacksonville Oregon. He watched as the Indian went into the cave and a few hours later came out with a bag of gold.

The trapper known as Wilson waited until the Indian had left the cave. He went into the cave it was dark and deep, he followed the tracks left by the Indian to a vein of gold.

This vein was as thick as a buggy wheel. It lined the whole length of the cave. When Wilson had chiseled out almost to much gold to carry. He started back to town. He made just outside of town when he saw the young Indian heading out of town back for the cave.

Wilson took out his gun and shot the young man. He thought of nothing but stopping anyone else from getting his gold.

The local Sheriff heard the gun shot and came riding up. Seeing the Indian shot dead and Wilson standing over him with his gun in hand. The Sheriff told Wilson to drop his gun he was under arrest for murdering an unarmed man.

The Sheriff told Wilson that he would have to stand trial for this crime. Wilson just laughed. A jury would never convict him of killing an Indian.

He would not be charged in town. But, he would be taken to the Indian camp to stand trial. Since the murder took place outside the city limits.

The Indian Chief was glad to see the murderer of one of his braves dealt with this way. He thanked the Sheriff and took Wilson to a TP for holding.

The tribe had a service of remembrance for the young brave. After the service they took Wilson to a tall tree. Wilson was going to get Indian justice and White man penalties. He was hanged by the neck until he was dead. His body was returned to the town and they could bury him.

But now for the clues as to where the cave is. When you leave Jacksonville head southwest towards the new Applegate Dam. On your way you will see a jagged rock on top of the mountain ridge on your right. A line of trees is now what you follow.

They look like they were planted just for you to follow. The ridge top is not far away if you look towards the beach and see a hollow. The caves are all along this jagged ridge.

Some will lead you to an open pit. But the one you want has an open sign it leads you down at a slight decline. The gold lies in the direction you came

Once you get to this spot you will really have no problem finding the way. It is clearly marked by old mother earth. No man made directions are here to worry about. The earth shows which way to go.

Good Luck to all I hope someone finds the pot of Gold.

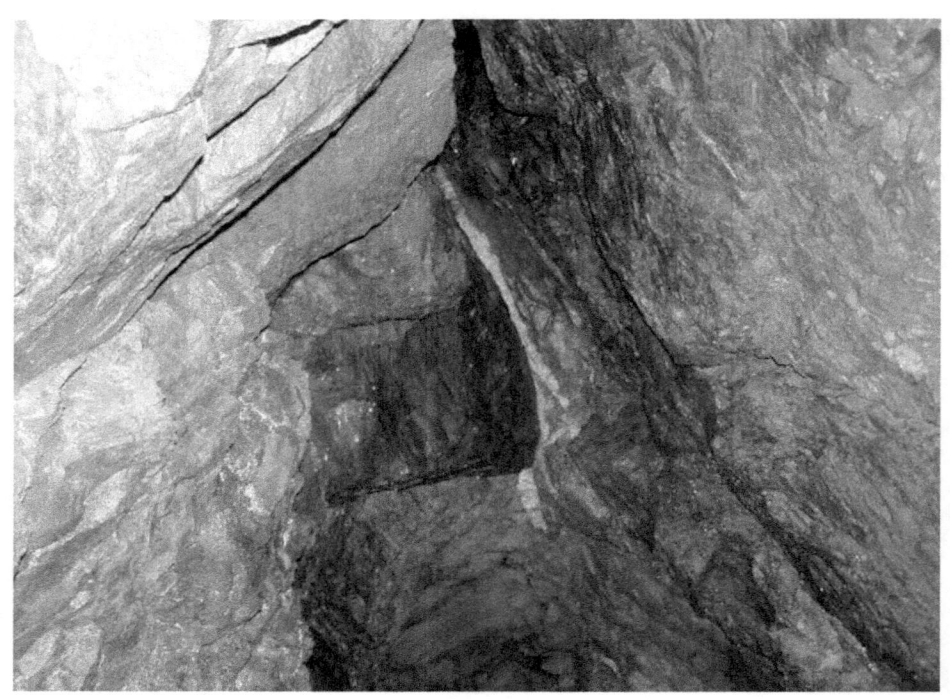

Shorty's List of Mines and Quarries
REFERENCE

This is how you read the following few pages. In the first line you see the name of the Mine or Quarry next you will see the map coordinates the Township (29S) Range (05W) Section (20) and section Fraction (NE).

Next, you see the GPS coordinates (43 –02-52N 123-18-12W). Next, where I could you will see the ore material mined or quarried (CR) and when it was possible to get the information you will see the Host Rock (SERPENTINITE HARZBURGITE).

I hope you have fun with this section. Be sure, do your own due diligence, and make sure your not crossing into private property or an active claim.

If you would like to see more of this list, you will have to order my CD LIST. It has more than 10,000 Mines and Quarries in the state of Oregon and is broke down by County. The following pages are for reference only. The list comes from State, County, Federal, BLM, DOGAMI, and my own records.

Areas covered in this book are Douglas, Coos, Curry, Josephine, and Jackson Counties.

Contents: Page

In Douglas County

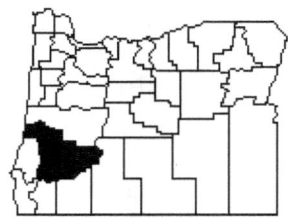

NAME

TOWNSHIP RANGE   SECTION   SECTION FRAC

NORTH LATITUDE   WEST LONGITUDE

MIND MATERIAL

A MINE
29S   05W   20   NE
43-02-52N   123-18-12W
CR   SERPENTINITE HARZBURGITE

ABEENE
25S   04W   18   CENTER
43-23-47N   123-12-58W

AGEE BAR PIT & PLANT
22S   07W   20
43-38-20N   123-33-30W
SAND & GRAVEL

"ALFRED SHIRLEY, JR./MONTOYA SITE"
28S   06W   11
43-08-48N   123-23-53W
SAND & GRAVEL

ALLEN PROSPECT
23S   04W   21
43-33-16N   123-10-53W
BASALT

ANDERSON
29S   06W   11
43-04-03N   123-22-05W
SAND & GRAVEL

ANDRAUEFF MEAOWS
29S   01E   1
43-04-21N   122-39-15W
SAND & GRAVEL

B AND H
29S   01W   13   "NE, NW"
43-03-40N   122-46-33W
VOLCANICS

BALDWIN
29S   08W   7
43-03-48N
123-41-25W

BALDY QUARRY
22S   04W   31
43-37-10N   123-12-50W
STONE

111

BANFIELD
        31S   02W   34
        42-50-07N   122-55-53W
        CU AG SCHISTOSE GREENSTONE

BARTON ROAD PIT
        32S   05W   30
        42-46-06N   123-19-58W
        SAND & GRAVEL

BASALT QUARRY
        22S   05W   21   NW
        43-38-41N   123-18-04W
        STONE (BASALT)

BASALT QUARRY
        24S   05W   11   NE
        43-30-11N   123-15-02W
        STONE (BASALT)

BASALT QUARRY
        27S   06W   23
        43-12-27N   123-22-10W
        STONE (BASALT)

BASALT QUARRY
        27S   05W   12   SE
        43-13-57N   123-13-52W
        STONE (BASALT)

BASALT QUARRY
        32S   04W   5
        42-49-14N   123-12-04W
        STONE (BASALT)

BEAR CREEK QUARRY
        22S   06W   24
        43-38-38N   123-21-02W
        STONE (BASALT)

BEAR CREEK QUARRY
        26S   08W   1
        43-20-23N   123-35-29W
        STONE

BEAVER SPRING
        31S   05W   7   "SW, SW, SW"

42-53-01N   123-20-49W
METAVOLCANICE

BECKLEY BAR
    30S   05W   18   SE
    42-57-30N   123-19-40W
    SAND & GRAVEL

BEN HUR
    29S   04W   35
    43-00-35N   123-08-54W
    METAGABBRO

BIG BALDY QUARRY
    23S   04W   1
    43-36-02N   123-07-24W
    STONE

BIG TOM FOLLEY
    22S   07W   2
    43-41-21N   123-29-20W
    SAND & GRAVEL

BLACK BOY
    33S   04W   5
    42-43-56N   123-12-25W
    CR   SERPENTINITE

BLUE HOLE
    23S   06W   6
    43-35-44N   123-26-49W
    SAND & GRAVEL

BOGGS SITE
    28S   05W   6   SW
    43-09-36N   123-20-26W
    STONE (BASALT)

BOLON ISLAND
    21S   12W   26   SW
    43-42-53N   124-06-00W
    SAND & GRAVEL

BONANZA MINE
    25S   04W   16   SW
    43-23-40N   123-10-57W

HG   "TUFFACEOUS   SANDSTONE,   ARKOSIC   SANDSTONE, SILTSTONE"

BOULDER DIVIDE ROAD NO.2844
     29S   01W   22
     43-02-45N   122-48-27W

BROSI PIT
     28S   06W   22
     43-07-35N   123-23-38W
     SAND & GRAVEL

BROWN QUARRY USFS
     30S   02W   33
     42-55-38N   122-57-07W
     STONE

BROWNIE QUARRY & PLANT
     30S   02W   33
     42-55-38N   122-57-07W
     STONE

BRUSHY BUTTE
     28S   04W   18
     43-05-50N   123-11-48W
     TUFFACEOUS SANDSTONE

BUCK FORK PLACERS
     28S   04W   12
     43-06-51N   123-07-31W
     STREAM AND BENCH
     GRAVEL;   BEDROCK-DECOMP0SED DIORITE

BUCKHORN
     30S   07W   1
     42-59-43N   123-28-30W
     CR   SERPENTINITE

BUCKHORN LIMESTONE
     27S   04W   14
     43-13-22N   123-08-28W

BUENA VISTA
     29S   02W   34
     43-01-03N   122-55-55W
     HG   ANDESITE FLOWS ADN TUFF BRECCIA

BULL RUN
  32S 05W 30 "S, S"
  42-45-42N 123-13-39W
  SCHIST
BULL RUN PLACERS
  32S 05W 25
  42-45-19N 123-14-28W
  AU

BUTTE PROSPECT
  24S 04W 26 CENTER
  43-27-17N 123-08-14W
  TUFFACEOUS SANDSTONE

BUZZARD BAY BAR
  22S 07W 30
  43-37-59N 123-34-07W   SAND & GRAVEL

BYRON LIMESTONE
  29S 07W 5
  43-04-36N 123-32-33W

C B RANCH PIT
  28S 06W 5
  43-09-40N 123-25-38W
  STONE

"CABIN CREEK, MARSH CREEK #75 QUARRY # 2"
  24S 05W 23
  43-28-04N 123-15-19W
  STONE

CALLAHAN MINE
  27S 07W 14 CENTER
  43-08-06N 123-29-31W
  COAL SANDSTONE AND SILTSTONE

CAMAS CREEK PIT
  27S 03E 15
  43-13-17N 122-26-22W
  STONE

CAMAS VALLEY AREA COAL
  29S 08W 22
  43-01-28N 123-37-33W
  "SANDSTONE",

CASEBEER QUARRY
        26S   03W   31
        43-16-06N   123-05-41W
        STONE

CASTEELS MINES COMPANY
        28S   04W   14 "SW, SW, SW"
        43-07-56N   123-09-51W
        STREAM GRAVEL

CEDAR SPRINGS PROSPECT
        32S   04W   25
        42-45-41N   123-06-54W
        SERPENTINTE

CHANDLER SITE
        28S   06W   10   SE
        43-08-43N   123-23-13W
        SAND & GRAVEL (TOPSOIL)

CHANEY
        28S   04W   19   S
        43-05-49N   123-11-48W

CHARLOTTE CREEK QUARRY
        22S   10W   17   NW
        43-39-34N   123-55-28W
        STONE

CHIEFTAIN
        29S   03W   20   NW
        43-02-27N   123-05-20W
        METAGABBRO

CHILCOOT GROUP
        25S   01E   8
        43-25-00N   122-42-44W
        STONE

CHRISTENSEN
        28S   06W   4
        43-09-59N   123-24-47W
        STONE (BASALT)

CLEGHORN
        20S   07W   32   "SE, SE"
        42-54-52N   123-32-54W
        STONE (SANDSTONE)

CLOUDY DAY
    32S   04W   33   N
    42-45-06N   123-10-47W
    AU AG GREENSTONE

CODY-LONG QUARRY AND PLANT
    23S   05W   15
    43-33-56N   123-16-10W
    SAND & GRAVEL

COFFEE CREEK
    30S   02W   6
    42-59-44N   122-58-53W

COMPTON QUARRY
    27S   05W   15
    43-13-30N   123-16-59W
    STONE (BASALT)

COMSTOCK COAL
    21S   04W   16   "N, NW"
    43-44-57N   123-10-60W
    SHALE AND SANDSTONE

CONTINENTAL MINE
    29S   03W   20   NW
    43-02-26N   123-05-45W
    METAGABBRO

COOPER AND YOUNG PROSPECT
    31S   02W   3   "SW, SW, SW"
    42-54-05N   122-56-27W
    PERIDOTITE

COPPER BUTTE PROSPECT
    32S   02W   9
    42-48-13N   122-56-21W
    SCHIST
COW CREEK
    31S   07W   1
    42-54-38N   123-28-52W
    SAND & GRAVEL

CRUMPTON
    30S   01W   4
    42-59-47N   122-49-50W

CUNNINGHAM QUARRY
    28S   06W   1

```
        43-09-35N       123-21-21W
        SAND & GRAVEL

CUNNINGHAM QUARRY
        28S     06W     1       SW
        43-09-35N       123-21-22W
        STONE

CURTIS
        33S     04W     5       W
        `42-43-45N      123-12-13W
        SERPENTINTE

DADS CREEK QUARRY
        32S     07W     21
        42-46-18N       123-32-12W
        STONE

DAN PARKER
        28S     06W     9
        43-08-46N       123-24-44W
        SAND & GRAVEL

DAN PARKER PIT
        27S     05W     18
        43-13-10N       123-20-09W
        STONE (BASALT)

DAYS CREEK BAR AND PLANT
        30S     04W     16
        42-57-31N       123-10-19W
        SAND & GRAVEL
DEB'S PIT
        22S     06W     8       SE
        43-40-05N       123-26-03W
        STONE

DIAMOND BUTTE
        32S     07W     33
        42-45-07N       123-31-38W
        AU

DIECKMAN PLACER
        28S     04W     13
        43-08-08N       123-07-23W
        AU

DILLARD QUARRY
        28S     06W     32      "NE, NE"
```

43-05-48N   123-25-25W
STONE

DING

26S   06W   19
43-17-22N   123-27-26W
SAND & GRAVEL

DODSON BUTTE PROSPECTS

28S   04W   19
43-07-51N   123-12-39W
SERPENTINTE

DODSON DEPOSIT

28S   05W   14   SW
43-02-46N   123-15-55W
STONE (LIMESTONE)   LIMESTONE

DOOLEY PROSPECT

29S   03W   17
43-02-53N   123-05-48W

DOUGLAS

32S   07W   27   NW
42-45-51N   123-30-59W
AU   GREENSTONE

DREW

31S   02W   13
42-52-45N   122-53-31W
BASALT AND TUFF

DREW PROSPECT

30S   02W   13
42-58-17N   122-53-50W

E.H. ROSEBOROUGH CLAIM

29S   01W   13
43-03-24N   122-24-17W

EAST FORK PIT

25S   02W   11
43-24-12N   122-53-56W
STONE

ELK VALLEY QUARRY

22S   05W   14
43-39-09N   123-15-10W
STONE

ELKHEAD MINE
        23S   04W   21   "NE, NE, NE"
        43-33-33N   123-10-10W
        HG   TUFFACEOUS

EMMETT L. PHILLIPS
        27S   05W   35
        43-10-59N   123-15-05W
        STONE

EVANS GROUP
        32S   02W   5"S, S, S, CENTER"
        42-49-10N   122-58-12W

FAIRVIEW PIT
        25S   02W   16
        43-23-22N   122-56-28W
        STONE

FISHER BAR
        28S   06W   21
        43-06-59N   123-25-11W
        SAND & GRAVEL

FISHER PLACER
        29S 07W   22
        43-01-52N   123-31-12W
        AU

FISHER PROPERTY
        28S   05W   30   "E, SW"
        43-06-06N   123-20-10W
        LIMESTONE

FLAT CLAIM
        32S   02W   1   CENTER
        42-49-10N   122-52-55W

FLYING SQUIRREL
        33S   04W   7
        42-43-05N   123-13-06W

FORGET-ME-NOT
        33S   05W   12   "SW, NE"
        42-43-11N   123-14-08W
        AU AG GREENSTONE

FORT KNOX

28S   04W   28
43-06-38N   123-31-05W

FOSTER CREEK GOLD PROSPECTS
29S   03E   23   "SE, SE, SE"
43-01-35N   122-24-21W
"TUFF, ANDESITE"

FOUR POINT   33S   05W   1
42-43-33N   123-14-52W
CR   SERPENTINTE

FRISTOE
31S   05W   11
42-58-41N   123-15-22W
FE

FROZEN CREEK MINE
28S   04W   19   "SE, SE"
43-06-48N   123-12-26W
CR   SERPENTINTE

GALE CREEK DAM
32S   04W   2
42-49-18N   123-08-42W
SAND & GRAVEL

GALESVILLE
31S   04W   28   "SW, NW"
42-51-01N   123-11-13W
STONE

GLUCINUM
26S   03E   34
43-15-45N   122-26-53W

GOLD BLUFF MINE
31S   05W   5   S
42-54-16N   123-18-60W
AU   GREENSTONE

GOLD CUT   30S
02W   27   NE
42-55-37N   122-55-35W
AU   GNEISSIC/ AMPHIBOLITE

GOLD MINE

29S 07W 20
43-02-15N 123-33-30W
AU

GOLD MINE
32S 07W 20
42-46-28N 123-33-25W
AU

GOLD MINE
32S 06W 17 "NE, NW"
42-47-45N 123-26-18W
AU

GOLD MINE
32S 06W 17 "SE, NW"
42-47-31N 123-26-18W
AU

GOLD MINE
32S 04W 29 SE
42-45-31N 123-11-38W
AU

GOLD PROSPECT
30S 02W 7
42-59-00N 122-59-40W

GOLDEN GATE
31S 06W 23
42-51-43N 123-23-07W
AU CU AG CHLORITE SCHIST

GOODIN PIT
29S 06W 5 "NW, NW"
43-05-00N 123-26-24W
STONE

GOOLAWAY QUARRY & PLANT
30S 02W 33
42-55-38N 122-57-07W
STONE

GOPHER
29S 01W 11 CENTER
43-04-12N 122-47-35W
RHYOLITE

GRAY BOY
        33S   04W   5   "SW, NE"
        42-43-55N   123-11-49W
        CR   DUNITE

GRAYBACK MOUNTAIN PROSPECT
        32S   07W   13
        42-47-20N   123-28-29W
        METAVOLCANICS CHLORITE SCHIST

GREEN COPPER
        30S   07W   35
        42-55-39N   123-30-34W

GREEN MOUNTAIN
        32S   04W   34   SW
        42-44-25N   123-10-02W
        SERPENTINTE

GREEN MOUNTAIN PROSPECT
        32S   04W   27   "SE, SE, SE"
        42-45-21N   123-10-14W
        AU AG GREENSTONE

GREEN QUARRY
        28S   06W   2
        43-09-43N   123-22-24W
        SAND & GRAVEL

GREEN VALLEY DEPOSIT
        24S   06W   21
        43-27-55N   123-25-00W
        STONE (LIMESTONE)   CALCAREOUS SHALE

GREEN VALLEY QUARRY
        24S   06W   22
        43-28-15N   123-24-41W
        STONE

HACKLER & HEIGHTS PLACER
        30S   05W   32
        42-54-49N   123-26-31W
        AU

HACKLER HEIGHT PLACER
        30S   06W   32   S
        42-54-48N   123-25-54W

AU   BENCH GRAVELS

HALES PROPERTY
        22S   10W   17
        43-39-32N   123-54-52W
        SAND & GRAVEL

HALES PROPERTY
        22S   10W   17
        43-39-32N   123-54-53W
        SAND & GRAVEL

HAMLIN IRON PROSPECT
        31S   02W   20
        42-52-14N   122-58-40W

HAMLIN PRAIRIE
        31S   01W   9
        42-53-27N   122-50-01W
        SAND & GRAVEL

HARKINS
        31S   02W   22   CENTER
        42-51-52N   122-55-49W
        SCHIST AND AMPHIBOLITE

HARRINGTON DEPOSIT
        28S   05W   21   "SW, SW"
        43-06-57N   123-18-02W
        LIMESTONE

HATFIELD LIMESTONE
        27S   04W   31
        43-10-47N   123-12-30W
        STONE

HOGAN SITE
        24S   05W   11
        43-30-10N   123-15-03W
        STONE (BASALT)

HOGUM CREEK
        32S   04W   20
        42-46-15N   123-11-02W
        AU   CHANNEL GRAVELS

HOLMS
        26S   02W   22   SW
        43-17-20N   122-56-49W

124

STONE (BASALT)

HUCKLEBERRY MINE
    31S   05W   7   CENTER
    42-53-28N   123-20-10W
    AU AG CHLORITIC SCHIST GREENSTONE

INDIANA
    32S   05W   25
    42-45-28N   123-14-54W

J. L. PROSPECT
    29S   01W   17   NW
    43-03-30N   122-51-25W
    BASALT AND TUFF BRECCIA

JAMES R.
    29S   02W   34   "E, E, E, CENTER"
    -00-45N 122-55-17W
    ANDESITE LAVAS AND TUFF

JEW GULCH PLACER
    28S   04W   11
    43-08-49N   123-08-12W
    AU

JOB #5044
    28S   06W   16
    43-07-58N   123-24-36W
    SAND & GRAVEL

JOHH HALL MINE
    29S   03W   19
    43-02-21N   123-05-58W
    METAGABBRO

JOHNNY CREEK
    22S   06W   2   NW
    43-41-26N   123-23-02W
    STONE

JONES BAR
    23S   07W   9
    43-35-18N   123-31-36W
    SAND & GRAVEL

JUDD CREEK
    30S   06W   2   SW
    42-59-14N   123-22-48W
    STONE
KELLER
    26S   06W   36   SW
    43-15-39N   123-21-16W
    STONE

KESTER ROAD UPPER QUARRY
    27S   05W   10
    43-13-55N   123-16-53W
    STONE

KOKO LANE SHALE PIT
    30S   04W   19   SW
    42-56-47N   123-13-14W
    STONE (SHALE ROCK)

KUMMER BAR
    30S   05W   18
    42-57-43N   123-20-07W
    SAND & GRAVEL

LAKE CREEK QUARRY
    23S   10W   24
    43-34-29N   123-48-29W
    STONE

LAST CHANCE PROSPECT
    32S   06W   5   SW
    42-49-11N   123-26-38W
    GREENSTONE

LAST CHANCE SULFUR MINE
    29S   03E   2   "SW, SW, SW"
    43-04-10N   122-26-24W
    ANDESITE AND/OR BASALT AND TUFF

LAURANCE
    29S   07W   18   SW
    43-03-01N   123-34-21W
    STONE (BASALT)
LAUREL
    28S   01E   24   "SW, CENTER"
    43-06-56N   122-38-59W
    RHYOLITE DIKE OR PLUG

LEES CREEK LODE

28S   04W   15
43-06-00N   123-08-00W
AU

LEVAN'S LEDGE
31S   05W   5   NE
42-54-38N   123-18-43W
AU   GREENSTONE

LINDA MARIE
30S   07W   34
42-55-06N   123-30-41W
CR   SERPENTINITE

LITTLE VALLEY PIT
27S   05W   34
43-10-19N   123-16-28W
SAND & GRAVEL

LONGBRAKE
25S   04W   20
43-23-10N   123-11-36W
UMPQUA SANDSTONE

LONGRIDGE QUARRY
27S   08W   1
43-15-03N   123-35-35W
STONE

LOOKING GLASS CR.
29S   07W   4
43-04-26N   123-32-02W
SAND & GRAVEL

LOOKINGGLASS QUARRY
27S   06W   29
43-11-21N   123-26-18W
SAND & GRAVEL

LOST FORTY
31S   02W   21
42-52-02N   122-57-02W

LOUGHS MINE
26S   01E   13
43-18-25N   122-38-53W
TUFF AND LAVAS

L-P QUARRY
29S   06W   1   NW

        43-04-56N   123-21-33W
        STONE

LUCKY CUSS CLAIMS
        30S   02W   31
        42-55-28N   122-59-25W
        SCHIST

"LUCKY CUSS NOS. 1, 2, 3, 4"
        31S   01W   24 CENTER
        42-51-54N   122-46-09W

LUCKY NINE GROUP
        30S   07W   25   "S, S, S"
        42-55-42N   123-28-31W
        CR   SERPENTINITE
LUCKY STRIKE
        31S   02W   10   CENTER
        42-53-41N   122-55-53W

LUCKY STRIKE NO. 1
        24S   04W   25   CENTER
        43-27-17N   123-07-03W

MANNING
        25S   04W   2   "W, W, W, CENTER"
        43-25-31N   123-08-51W

MARTIN PLACER
        28S   04W   15
        43-07-56N   123-08-58W   AU

MAUD S.
        29S   02W   34   "N, CENTER"
        43-00-54N   122-55-53W
        HG   ANDESITE FLOWS AND TUFFS

MAUPIN BAR
        23S   07W   10
        43-34-47N   123-31-16W
        SAND & GRAVEL

MAY
        30S   02W   27
        42-56-23N   122-55-46W

MCDOUGAL QUARRY
        22S 05W 19
        43-38-41N   123-20-30W

STONE

MELODY PROSPECT
    29S   06W   31
    43-00-42N   123-25-46W
    "BASALT, DIORITE"

MILDRED
    32S   04W   33   NW
    42-45-11N   123-11-14W
    AU   GREENSTONE

MILLER
    31S   04W   13
    42-53-00N   123-07-20W

MILLS PROSPECT
    29S   01W   15   CENTER
    43-03-20N   122-48-48W
    ANDESITE AND TUFFS

MILLTOWN/GILMER
    23S   04W   5
    43-35-44N   123-11-23W
    STONE (BASALT)

MISER
    32S   04W   29
    42-45-51N   123-11-56W

MORGAN QUARRY
    27S   06W   2
    43-15-01N   123-22-44W
    STONE (BASALT)

MORGAN SITE
    26S   06W   36   SW
    43-15-39N   123-21-43W
    STONE

MORRIS
    31S   04W   6
    42-54-32N   123-13-03W

MOUNT OBETH
    28S   06W   32
    43-05-43N   123-25-56W
    STONE

MOWICH CREEK PIT
      26S   03W   9
      43-19-31N   123-03-09W
      SAND & GRAVEL

MT BALDY QUARRY
      22S   05W   28
      43-37-56N   123-18-09W
      STONE

MT NEBO QUARRY
      27S   06W   25
      43-12-19N   123-21-39W
      SAND & GRAVEL

MYRNA - THELMA NOS. 1 & 2
      29S 01W   17 CENTER
      43-03-21N   122-51-07W

NEWTON CREEK PIT
      27S   05W   5
      43-15-25N   123-19-05W
      STONE

NEWTON CREEK QUARRY
      26S   05W   32
      43-16-03N   123-19-00W
      STONE (BASALT)

NICKEL MOUNTAIN
      30S   06W   17   CENTER
      42-57-38N   123-26-26W
      NI CO PGM LATERITE

NICKEL MOUNTAIN GROUP
      30S   06W   17
      42-57-37N   123-26-37W
      NI   "SAPROLITE,"

NICKEL MOUNTAIN GROUP
      30S   06W   20
      42-57-11N   123-26-07W
      CR

NICKLE WASTE DUMP
      30S   06W   28
      42-55-50N   123-24-45W
      STONE

NINEMILE QUARRY & PLANT
        33S   06W   4
        42-44-05N   123-25-06W
        STONE

NIVINSON
        32S   02W   16
        42-47-05N   122-56-29W
        AMPHIBOLITE SCHIST MICA SCHIST

NO NAME QUARRY
        23S   05W   12
        43-35-23N   123-14-34W
        STONE

NONPAREIL MINE
        25S   04W   3   "SW, SW"
        43-24-49N   123-09-47W
        HG   TUFFACEOUS SANDSTONE

NORTH UMPQUA-LITTLE RIVER AREA
        27S   03W   10
        43-15-53N   123-04-26W
        "CONGLOMERATE, SANDSTONE"

NR MILL CREEK
        22S   10W   28   NE
        43-38-07N   123-52-00W
        STONE (SANDSTONE)

ODEN HATFIELD DEPOSIT
        27S   04W   33
        43-11-01N   123-10-58W
        STONE

OLLALA PLACERS
        29S   07W   22
        43-02-05N   123-32-44W
        AU AG STREAM GRAVELS

OLLIVANT'S PIT
        28S   07W   12
        43-08-53N   123-28-35W
        STONE

OREGON M & P CO.

```
          32S     07W     28
          42-45-22N       123-31-36W
          AU PT   HIGH BENCH GRAVEL
```

OREGON PORTLAND CEMENT CO
```
          28S     05W     20
          43-07-01N       123-18-34W
          STONE (LIMESTONE)
```

OREGON PORTLAND CEMENT CO. QUARRY
```
          28S     05W     29
          43-06-53N       123-18-26W
          STONE (LIMESTONE)   LIMESTONE
```

OVERMAN LODE
```
          28S     04W     14
          43-07-59N       123-08-18W
```

OVERMAN PLACER
```
          28S     04W     15
          43-07-38N       123-09-19W
          AU
```

PAROZ PIT
```
          27S     06W     32
          43-10-25N       123-26-03W
          STONE
```

PEEL PROSPECTS
```
          26S     03W     27 "S,S,S, CENTER"
          43-21-25N       122-55-02W
          SERPENTINITE HARZBURGITE
```

PENNEL AND FARMER PROSPECT
```
          30S     02W     27
          42-56-14N       122-56-10W
          AMPHIBOLITE SCHIST AND GNEISS
```

POOR BOY
```
          29S     01W     16
          43-03-06N       122-49-54W
          ANDESITE FLOWS TUFF AND AGGLOMERATE
```

PROSPECT
```
          28S     03E     27
          43-06-40N       122-26-45W
          "TUFF, ANDESITE"
```

PUZZER
        32S   04W   34
        42-44-47N   123-09-59W
        AU   SHALE

QUARTZ MOUNTAIN SILICA
        28S   01E   2
        43-09-36N   122-40-25W
        SILICA (LODE) SILICIFIED RHYOLITE TUFF

QUARTZMILL
        33S   05W   1   NW
        42-44-15N   123-14-46W
        GREENSTONE

RAELYNNE PROSPECT
        30S   02W   34
        42-55-16N   122-55-58W
RAINVILLE
        31S   02W   9
        42-53-43N   122-56-55W

RAINY DAY
        30S   04W   15
        42-57-28N   123-09-31W
        CR   SERPENTINITE

RALF N HAKANSON
        24S   05W   33
        43-26-37N   123-15-10W
        SAND & GRAVEL

RALLS PLACER
        29S   04E   19   SW
        43-01-35N   122-23-42W
        AU

RED CLOUD MINE
        32S   02W   21
        42-46-43N   122-56-41W
        HG   AMPHIBOLITE
        QUARTZ-MICA-HORNBLENDE SCHIST

RED HILL NO. 1
        32S   03W   6   "S, S, S, CENTER"
        42-48-37N   123-05-57W

RED HILL PROSPECT
    32S   02W   5 "W, W, W, CENTER"
    42-49-36N   122-58-11W
    TALC SCHIST
REDING SITE
    29S   07W   4
    43-05-10N   123-32-10W
    STONE

REEDSPORT YARD
    21S   12W   35
    43-42-11N   124-05-22W
    STONE

ROMAN NOSE
    19S   09W   14   SE
    43-54-37N   123-43-40W
    STONE
ROSS CLAIMS
    29S   01W   17   "W, W, W, CENTER"
    43-03-19N   122-51-43W
    ANDESITE FLOW AND TUFF

ROUND PRARIE PIT AND PLANT
    28S   06W   35
    43-05-25N   123-22-35W
    SAND & GRAVEL

ROWE PROSPECT
    30S   04W   16   CENTER
    42-57-46N   123-10-40W
    BASALT

ROWLEY MINE
    32S   02W   4   S
    42-48-57N   122-56-34W
    CHLORITE SCHIST AMPHIBOLITE

RUMMEL BAR
    28S   06W   29
    43-06-33N   123-26-05W
    SAND & GRAVEL

SAND PIT
    21S   12W   18   NE
    43-44-58N   124-09-56W
    SAND & GRAVEL (SAND)

SCIEFFELIN #2
29S   02W   6
42-59-54N   122-59-19W

SCOGGINS PIT AND PLANT
22S   10W   13
43-39-28N   123-51-24W
SAND & GRAVEL

SHADY PIT
27S   06W   36
43-10-33N   123-21-42W
SAND & GRAVEL

SHOO & YD
27S   06W   36
43-10-54N   123-21-39W
SAND & GRAVEL

SHORT CLAIM
31S   02W   21
42-51-41N   122-57-02W
ULTRAMAFICS

SHORT QUARRY
26S   05W   15
43-18-38N   123-17-04W
STONE

SILVER PEAK
31S   06W   26   "NW, NW"
42-50-57N   123-22-15W
CU ZN AU AG TUFFACEOUS GREENSTONE

SITTING DUCK
30S   03W   19
42-57-09N   123-06-34W
CHERT AND ARGILLITE

SLAUGHTER PEN PIT
30S   07W   34
42-54-55N   123-31-14W
STONE

SMITH BAR
30S   06W   26
42-56-10N   123-22-40W
SAND & GRAVEL

SMITH BAR
        30S   06W   33   NW
        42-55-33N   123-25-25W
        SAND & GRAVEL

SNOW CREEK QUARRY
        32S   03W   30
        42-45-28N   123-06-35W
        STONE
SNOWBIRD ROCK QUARRIES
        27S   01E   24
        43-12-34N   122-38-39W
        STONE

SPEEDWAY PIT
        27S   06W   35
        43-10-38N   123-22-21W
        STONE

STARVE OUT
        32S   04W   32
        42-44-56N   123-12-04W
        SERPENTINITE
STRAIGHT JOHN
        30S   02W   33
        42-55-38N   122-57-07W
        STONE

STROUBS MINE
        27S   02E   14   "S, S, S"
        43-12-48N   122-32-31W
        RHYOLITE

STUEMPGES PROSPECT
        29S   05W   16   "S, S, S"
        43-02-31N   123-17-58W
        SERPENTINITE

SULFIEDS MINE
        28S   04W   35   SW
        43-05-18N   123-08-34W
        SULFIDES

SUTHERLAND
        25S   04W   30
        43-22-02N   123-12-60W
        "TUFFACEOUS SANDSTONE, BASALT"

SWEETBRIAR
31S   05W   8
42-53-45N   123-19-29W
GREENSTONE

T AND M PROSPECT
32S   01W   11   "S, S"
42-47-55N   122-47-14W
TUFFS

TALLOW BUTTE PIT AND PLANT
29S   01W   34
43-01-01N   122-49-13W
SAND & GRAVEL

TEECO CORP/MELIUS QUARRY
28S   08W   23
43-07-00N   123-36-19W
STONE

TEN MILE PIT
28S   08W   23
43-07-33N   123-36-13W
SAND & GRAVEL

TEN MILE QUARRY #2
28S   08W   24
43-07-01N   123-35-58W
STONE

TENNESSE GULCH
33S   05W   2   "NW, NW"
42-44-06N   123-15-43W
AU   STREAM GRAVELS

THOMASON
32S   02W   16   SE
42-47-01N   122-56-09W
MICA SCHIST AMPHIBOLITE

THOMPSON
23S   04W   15   "SW, CENTER"
43-33-59N   123-09-43W

TWIN CEDARS PIT
24S   02W   14
43-29-04N   122-54-12W
STONE

UMPQUA COAL CO.
    23S   08W   21
    43-34-09N   123-38-56W
    "SANDSTONE, SILTSTONE"

UMPQUA PIT
    27S   06W   4
    43-15-15N   123-24-02W
    SAND & GRAVEL

UMPQUA SAND & GRAVEL PIT
    26S   05W   22
    43-17-56N   123-16-52W
    SAND & GRAVEL

UNION LEADER
    32S   05W   36
    42-45-04N   123-14-24W
    AU
UPDEGRAVE QUARRY
    26S   04W   29
    43-16-51N   123-12-04W
    SAND & GRAVEL

VICTORY
    32S   07W   33   NE
    42-44-52N   123-31-58W
    AU   HIGH BENCH GRAVELS

WARNER
    33S   04W   4
    42-44-01N   123-11-05W
    AU

WATSON BUTTE EMERY
    26S   05E   19
    43-17-30N   122-16-22W

WEAVER SITE
    28S   08W   23
    43-07-32N   123-36-13W
    STONE

WEYERHAUSER
    26S   03W   18
    43-18-26N   123-06-08W
    STONE

WHITEHORSE PLACER
     32S   04W   10   "S, S, S"
     42-47-50N   123-09-53W
     AU   CHANNEL GRAVELS

WILLIS
     33S   06W   9   "NW, NW"
     42-43-30N   123-25-24W
     METAVOLCANICS

WILMAR QUARRY
     29S   01E   19
     43-02-00N   122-44-34W
     STONE
WILSON PROSPECT
     23S   04W   8
     43-35-23N   123-11-47W

WILSON PROSPEST
     29S   01W   23
     43-35-23N   123-11-47W
     BASALT

WINCHESTER
     26S   06W   25   NW
     43-17-06N   123-21-47W
     SAND & GRAVEL

WINCHESTER BAY
     22S   12W   1
     43-41-44N   124-04-41W
     SAND & GRAVEL
WINSTON
     28S   06W   15
     43-07-41N   123-23-42W
     SAND & GRAVEL

WOLF RUNNER PIT
     27S   04W   14
     43-13-15N   123-07-51W
     STONE

WOOLEY
     19S   09W   16
     43-55-42N   123-46-10W
     STONE

YELLOW ROCK #1
        29S   01E   22
        43-01-50N   122-24-12W

YOUNGS BAR/NORDIC PIT
        26S   06W   18
        43-18-24N   123-27-12W
        SAND & GRAVEL

ZINC MINE
        29S   01W   23

        43-02-33N   122-47-40W

        TUFF BRECCIA

Coos County

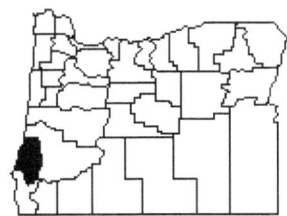

NAME
TOWNSHIP   RANGE   SECTION   SECTION FRACTION
NORTH LATITUDE   WEST LONGITUDE
MINED MATERIAL
Last Update of Claim or Quarry if I was able to get it

101 SAND AND GRAVEL PIT AND PLANT
       28S   14W   03
       43-10-34N   124-22-15W
       SAND & GRAVEL

4000 QUARRY
       25S   09W   06
       43-25-32N   123-48-36W
       STONE

AGATE BEACH CHROMIUM OCCURENCE
       27S   14W   04   "SW,NW"
       43-15-36N   124-23-02W
       BLACK SANDS

AIRPORT PIT   C
       25S   13W   09
       43-25-02N   124-14-57W
       SAND & GRAVEL(SAND)

ALBEE
       29S   13W   04
       43-05-35N   124-14-58W
       COAL

ALBERTSON SITE   C
       28S   13W   31
       43-06-45N   124-17-25W
       SAND & GRAVEL(TOPSOIL)

ALL COAST CONCRETE
       30S   12W   04
       42-59-33N   124-08-51W
       SAND & GRAVEL

ALLEN
       27S   14W   21
       43-13-05N   124-22-02W

ALLMAN (HARRY LAWSON)   C
       28S   14W   29
       43-06-49N   124-23-27W
       SAND & GRAVEL

ANACONDA
       33S   12W   33
       42-40-45N   124-07-59W

ARAGO SAND & GRAVEL PIT
       28S   13W   36   NE
       43-06-17N   124-11-11W
       SAND & GRAVEL   GRAVEL

ARCHER & SENGSTACKEN NO 1
  25S 13W 36
  43-21-27N   124-11-24W
  COAL
ARCHER & SENGSTACKEN NO 2
  25S 13W 35
  43-21-37N   124-11-55W

ARCHER & SENGSTACKEN NO 3
  25S 12W 36
  43-21-40N   124-10-45W
  COAL
ARCHER & SENGSTACKEN NO 4
  25S 12W 31
  43-21-28N   124-10-13W
ART E JONES
  28S 12W 10 "NW,SE"
  43-09-42N   124-06-14W
  STONE
AZALEA LAKE ROAD PIT
  31S 12W 13
  42-53-00N   124-04-26W
  STONE
BANDON
  28S 14W 31
  43-06-33N   124-23-56W
BANDON PLACER
  29S 15W 12 NE
  43-04-41N   124-25-07W
  TI BEACH SANDS
BANDON PLANT (SECTION 8 PIT) C
  28S 14W 08 "SW,NE"
  43-09-51N   124-22-41W
  SAND & GRAVEL GRAVEL

BASALT QUARRY
  26S 12W 24 SW
  43-17-53N   124-04-33W
  STONE(BASALT)
BASALT QUARRY
  26S 12W 24 NE
  43-18-21N   124-03-55W
  STONE(BASALT)

BAY AREA HOSPITAL
  25S 13W 22

             43-23-16N   124-13-44W
             SAND & GRAVEL
BEACH CHROMIUM OCCURRENCE
             28S   15W   36   "S,S"
             43-05-47N   124-25-52W
             BLACK SANDS
BEAVER HILL
             27S   13W   17
             43-13-59N   124-16-15W
             COAL   SANDSTONE AND SHALE
BELFAST
             26S   13W   02   SE
             43-20-22N   124-12-09W
             COAL   SANDSTONE AND SHALE
BIG CREEK
             26S   14W   16
             43-18-52N   124-21-56W   COAL

BIG SLIDE
             32S   12W   34
             42-45-14N   124-04-53W
             AU
BITUMINOUS COAL CO
             27S   13W   25
             43-11-57N   124-11-43W
             COAL
BLACK DIAMOND TUNNEL
             24S   13W   34
             43-27-04N   124-13-29W
             COAL
BLUE PIT
             32S   12W   27
             42-45-39N   124-05-19W
             SAND & GRAVEL
BLUE PIT & PLANT
             31S   12W   13
             42-52-56N   124-04-00W
             STONE
BLUEBIRD
             33S   12W   21
             42-21-12N   124-07-46W
BOLIVAR
             32S   10W   10
             42-47-46N   123-51-19W
             CU   BASALT-ANDESITE
BORROW PIT
             29S   14W   05   SE
             43-05-15N   124-22-42W
                                144

SAND & GRAVEL(FILL)

BORROW PIT
    29S   12W   05   "E,NE"
    43-05-23N   124-08-31W
    SAND & GRAVEL(FILL)
"BRINKS MINING, INC."   C
    26S   13W   36
    43-16-09N   124-11-00W
    SAND & GRAVEL COAL
BROADBENT
    30S   12W   08
    42-59-17N   124-09-25W
    SAND & GRAVEL
BROADBENT   C
    30S   12W   08
    42-59-01N   124-08-43W   STONE
BROWN SLOUGH
    26S   13W   31
    43-16-11N   124-16-57W
BULLARDS PIT & PLANT
    28S   14W   30
    43-07-03N   124-24-09W
    SAND & GRAVEL
BUNKER HILL
    25S   13W   35
    43-21-14N   124-12-11W
    COAL
BURCH
    25S   13W   34
    43-21-38N   124-13-32W
BUTLER
    31S   15W   27   "SW,SW"
    42-51-18N   124-28-10W   CR
    BLACK SANDS
CALEDONIA
    26S   13W   11
    43-19-44N   124-12-30W
    COAL   SANDSTONE AND SHALE
CARLSON NO 2
    25S   12W   07
    43-25-09N   124-10-44W
CEDAR POINT
    27S   13W   35
    43-10-58N   124-13-01W
CHARLESTON
    26S   14W   02
    43-20-58N   124-19-58W
    SAND & GRAVEL

CHICKAMIN MINE
    26S   14W   25   SE
    43-17-00N   124-18-17W
    BLACK SANDS
CHROME BEACH SAND OCCURENCE
    27S   14W   04
    43-15-49N   124-21-21W
    BLACK SANDS
CHROMIUM OCCURENCE
    26S   14W   13   "NE,NE"
    43-19-17N   124-17-59W
    BLACK SANDS
CHROMIUM OCCURENCE
    26S   14W   25   S
    43-17-01N   124-18-25W
    BROWN AND BLACK - STAINED SAND
CHROMIUM OCCURENCE
    26S   14W   36   NW
    43-17-21N   124-17-59W
    WELL CEMENTED BROWN AND
     BLACK STAINED SAND
CHROMIUM OCCURENCE
    27S   14W   01   NE
    43-15-46N   124-18-13W
    BLACK SANDS

CHROMIUM OCCURENCE
    27S   14W   08   "S,S"
    43-14-10N   124-23-26W
    BLACK SANDS
CHROMIUM PROSPECT
    28S   14W   33
    43-06-19N   124-21-55W
    CR
CLAY PIT   C
    25S   13W   28   NE
    43-22-42N   124-14-43W
    CLAY
COAL MINE
    27S   14W   01
    43-15-57N   124-18-31W   COAL

COAL OCCURRENCE
    26S   13W   36
    43-16-43N   124-11-49W
COAL PROSPECT

```
          25S     12W     19
          43-23-18N       124-09-31W
COALEDO MINES
          27S     13W     14      "SW,SW"
          43-13-20N       124-13-08W
          COAL   SANDSTONE AND SHALE


COALEDO SAND & GRAVEL
          27S     13W     15      NW
          43-14-08N       124-14-04W
          SAND & GRAVEL
COOPER BRIDGE PIT
          29S     12W     10      "NW,SE"
          43-04-16N       124-06-29W
          SAND & GRAVEL        GRAVEL
COOS COPPER
          32S     12W     29
          42-45-28N       124-07-45W
          GREENSTONE SEDIMENTS
COOSAND CORP.
          24S     13W     34
          43-26-24N       124-14-15W       SILICA SAND
COPPER KING
          33S     12W     33
          42-40-30N       124-08-08W
          SERPENTINE
COX CANYON CHROMIUM OCCURENCE
          26S     14W     36      "SW,SW"
          43-16-03N       124-18-44W
          FIRMLY CEMENTED BLACK AND BROWN STAINED SAND
CROOKED CREEK
          29S     14W     17      "NW,NW"
          43-03-59N       124-23-24W
          BLACK SANDS


DAVIS SLOUGH NO 1
          26S     13W     21
          43-17-59N       124-14-41W
DAVIS SLOUGH NO 2
          26S     13W     31
          43-16-37N       124-17-36W


DELMAR
          27S     13W     04      SE
          43-15-21N       124-14-18W
          COAL            SANDSTONE AND SHALE
DEVILS ROAD CHROMIUM OCCURENCE
          26S     14W     33      S
```

43-16-11N   124-21-54W
BLACK SANDS

DIVELBISS
        32S   12W   32
        42-45-14N   124-07-49W
        AU   SILTSTONE
DONALDSON NO 1
        28S   14W   12
        43-09-20N   124-18-06W
DONALDSON NO2
        28S   14W   12
        43-09-40N   124-18-00W

EAGLE MINE
        27S   14W   33   "N,NE"
        43-11-43N   124-21-43W
EAGLE MINE (COQUILLE)   C
        27S   14W   28   "NE,NE"
        43-11-52N   124-21-42W
        AU CR BLACK SANDS
EAST BLUFF-HEAD OF BROWN SLOUGH
        26S   14W   25
        43-16-54N   124-18-10W
        AU
ECKLEY QUARRY
        30S   12W   31
        42-55-59N   124-10-05W
        STONE
ECKLEY SITE
        30S   12W   29   "SW,NE"
        42-56-26N   124-08-55W
        STONE
EDEN RIDGE
        32S   11W   28
        42-45-55N   123-59-24W
        "SHALE, SANDSTONE"
EMPIRE MINES
        25S   13W   20   SW
        43-22-52N   124-16-09W
        COAL   SANDSTONE AND SHALE
ENCHANTED PRAIRIE C
        29S   11W   36   NW
        43-00-56N   123-57-17W   STONE
ENDICOTT CREEK PIT
        29S   12W   24   "W,SE"
        43-02-24N   124-04-06W
        SAND & GRAVEL   GRAVEL

ENEGREN
    25S   12W   27   "NW,SE"
    43-22-18N   124-06-02W   STONE

ENGLEWOOD
    25S   13W   34   "SW,SW"
    43-21-15N   124-13-54W
    COAL SANDSTONE AND SHALE
EUREKA
    28S   13W   08
    43-09-52N   124-15-55W   COAL
FAHY
    27S   13W   30
    43-12-29N   124-17-43W   COAL
FAIRVIEW SITE
    27S   12W   10   NE
    43-14-39N   124-06-06W   STONE(BASALT)

FAT ELK
    28S   13W   03   "SW,SW"
    43-10-15N   124-14-13W
    COAL   SANDSTONE AND SHALE
FAT ELK SITE   C
    28S   13W   03
    43-10-12N   124-14-08W
    STONE

FERBERISH
    27S   12W   18
    43-13-53N   124-10-10W

FITZGERALD
    33S   12W   26
    42-41-45N   124-06-04W
    CHERT

FIVE MILE POINT
    27S   14W   17   "SW,SW,SW"
    43-13-17N   124-23-41W
    BLACK SANDS
FLANAGAN
    25S   13W   27
    43-22-18N   124-14-06W
    COAL   SANDSTONE AND SHALE
FLETCHER MYERS PROPERTY
    27S   14W   16   SW
    43-13-37N   124-22-13W
    BLACK SANDS

FULLER QUICKSILVER
     32S   10W   10
     42-48-27N   123-51-13W
"GEIGER CREEK MINES, INC."
     28S 14W 32   "SE,SE"
     43-06-20N   124-22-39W   BLACK SANDS

GILBERTSON MINE
     25S   12W   06   NW
     43-26-13N   124-10-40W
     COAL   SANDSTONE AND SHALE
GLEN ALKEN CREEK SAND & GRAVEL
     28S   12W   18   SE
     43-08-22N   124-10-01W
     SAND & GRAVEL   GRAVEL
GOLD BACK- FULLER
     32S   10W   10
     42-48-33N   123-51-15W
     AU
GOLD PLACER
     31S   10W   36
     42-49-58N   123-49-49W
     AU
GRANITE CREEK PIT
     32S   12W   29
     42-45-44N   124-07-44W
     SAND & GRAVEL
GRAY CREEK PIT
     28S   12W   29   NE
     43-07-13N   124-08-44W
     SAND & GRAVEL   GRAVEL
GUERIN
     29S   12W   22
     43-02-42N   124-06-35W
GUNNELL NO 1
     26S   12W   06
     43-20-57N   124-09-34W   COAL
GUNNELL NO 2
     26S   12W   05
     43-20-48N   124-09-21W
HALL CREEK
     29S   13W   11
     43-04-37N   124-13-00W
HANSON
     24S   13W   13
     43-29-16N   124-11-26W   COAL
HAPPY HOOLIGAN
     28S   14W   36
     43-06-20N   124-18-26W

HARBOR TUG AND BARGE
    24S   13W   15
    43-29-04N   124-13-10W
    SAND & GRAVEL

HARDY
    25S   13W   01   "NE,SW"
    43-25-43N   124-11-28W
    COAL   SANDSTONE AND SHALE
HUNTLEY NO 1
    26S   12W   17
    43-19-12N   124-08-59W
    COAL
HUNTLEY NO 2
    26S   12W   17
    43-18-52N   124-09-29W   COAL
INDEPENDENCE
    33S   12W   23
    42-42-12N   124-05-29W
    CR AU   SERPENTINE
INDEPENDENCE MINE
    27S   14W   33   NE
    43-11-30N   124-22-15W

INDIAN CREEK PIT
    29S   12W   36
    43-00-40N   124-03-44W
    STONE
IOWA MINE
    28S   14W   33   "S,N"
    43-06-43N   124-22-09W
IRON MOUNTIAN
    33S   12W   13
    42-42-56N   124-05-15W

JOHN B CHROMIUM OCCURRENCE
    26S 14W 36   "SW,NE"
    43-16-41N   124-18-05W
    BLACK SANDS
JOHN R STINSON RANCH
    28S   10W   10
    43-08-51N   123-51-39W
JOHNSON CREEK CHROMIUM OCCURENCE
    29S   14W   09   NW
    43-04-51N   124-22-19W
    BLACK SANDS

JOHNSON CREEK PLACERS
     32S   12W   28
     42-45-41N   124-06-30W
     AU   STREAM GRAVELS
JOHNSON CREEK SAND & GRAVEL
     29S   14W   05   SW
     43-05-02N   124-23-18W
     MARINE BEACH AND OFFSHORE DEPOSITS

JOHNSON MT BPR PIT # 1
     32S   12W   11
     42-48-20N   124-04-07W
     SAND & GRAVEL
JOHNSON MT BPR PIT # 2
     32S   12W   10
     42-48-16N   124-05-23W
     SAND & GRAVEL
JUPITER GROUP
     33S   12W   11
     42-44-18N   124-06-25W
     META-ANDESITE METAGABBRO
KENDALL MINE
     27S   14W   08   NE
     43-14-44N   124-22-33W
     BLACK SANDS
KENROCK
     25S   12W   03
     43-26-03N   124-06-28W
     SAND & GRAVEL
KENSTONE   A
     24S   12W   26
     43-27-13N   124-05-13W
     SAND & GRAVEL
KENTUCK
     24S   12W   34   SE
     43-26-22N   124-06-36W
     STONE
KENTUCK SLOUGH
     25S   12W   03   N
     43-26-11N   124-06-41W
     STONE
KINCHELOE QUARRY
     29S   11W   36   "SW,NE"
     43-00-36N   123-57-29W
     STONE(BLUESHCHIST) MARINE BASALT
KLONDIKE
     27S   13W   18   SE
     43-13-35N   124-16-55W
     COAL   SANDSTONE AND SHALE

KOOSROCK QUARRY
  25S   12W   27   "N,NW"
  43-22-48N   124-06-44W
  STONE
KOOSTONE (BAKER QUARRY) A
  25S   12W   22   SE
  43-23-05N   124-06-22W
  STONE(BASALT)

LAKESIDE NO 1
  23S   12W   17
  43-34-20N   124-08-46W
  COAL
LAKESIDE NO 2
  23S   12W   20
  43-33-15N   124-09-38W
  COAL
LAMPA CREEK MINE
  28S   14W   25   "SE,SE"
  43-06-38N   124-17-50W
  COAL
LANE EXTENTION MINE
  27S   14W   33
  43-11-20N   124-21-50W
LAST CHANCE
  33S   12W   24
  42-41-60N   124-04-54W
  CR   SERPENTINE
LEEP
  33S   12W   24
  42-42-34N   124-05-17W
  CHERT
LEEP QUARRY
  28S   12W   32
  43-05-46N   124-08-52W   STONE

LIBBY
  26S   13W   03   N
  43-20-37N   124-14-46W
  COAL   SANDSTONE AND SHALE
LILLIAN
  25S   12W   04
  43-21-11N   124-07-50W
  COAL   SANDSTONE AND SHALE
LILLIAN BLACK DIAMOND
  26S   12W   04
  43-21-02N   124-07-42W   COAL

LITTER JUPITER

```
        33S     12W     11
        42-44-06N       124-06-00W
LITTLE LODE
        33S     12W     15
        42-43-17N       124-07-04W
LONE ROCK NO 1
        24S     13W     25
        43-27-24N       124-11-47W
        COAL
LONE ROCK NO 2
        24S     13W     36
        43-27-01N       124-11-36W
        COAL

LYONS
        28S     13W     19      NE
        43-08-08N       124-17-01W
        COAL    SANDSTONE AND SHALE

LYONS (HARLOCKER HILL)
        28S     13W     23
        43-07-44N       124-12-06W
        COAL
MAGNA BONUS
        32S     12W     29
        42-45-39N       124-07-36W
        "DACITE PORPHYRY, GABBRO"
MAHAFFY QUARRY
        25S     12W     13      NE
        43-24-24N       124-03-52W              STONE(SANDSTONE)
MAIN QUARRY AND PLANT
        25S     13W     34
        43-21-30N       124-13-35W
        STONE
MARSTARS
        28S     13W     21
        43-08-04N       124-14-32W

MARTIN MINE
        27S     13W     09      SE
        43-14-29N       124-14-45W
        COAL    SANDSTONE AND SHALE
MATHEWS M J
        26S     14W     26
        43-16-52N       124-19-45W
MAXWELL
        26S     13W     27      SW
        43-17-04N       124-13-33W
        COAL    SANDSTONE AND SHALE
```

MC ADAMS
     30S   14W   20
     42-57-30N   124-23-12W
MC CLAIN MINE
     27S   13W   35   SE
     43-11-11N   124-12-17W
     COAL   SANDSTONE AND SHALE
MCLEOD QUARRY   C
     28S   12W   30   NW
     43-07-09N   124-10-14W
     STONE(BASALT)
MCMULLEN PIT
     27S   13W   36
     43-11-08N   124-10-53W
     STONE
MCMULLEN PIT   C
     29S   11W   20   SW
     43-02-12N   124-02-15W
     STONE
MERCHANTS BEACH
     27S   14W   17
     43-14-10N   124-23-26W
MERCHEN & REED GRAVEL CO (1)
     31S   12W   24   "NW,NW"
     42-52-25N   124-04-45W
     SAND & GRAVEL   GRAVEL
MERCHEN AND REED GRAVEL CO (2)
     30S   12W   08
     42-59-04N   124-09-02W
     SAND & GRAVEL
MERCHEN AND REED GRAVEL CO (3)
     30S   12W   15
     42-57-48N   124-06-36W   SAND & GRAVEL

MESSERIE
     26S   12W   05
     43-20-26N   124-09-02W
MESSERLE SITE   C
     27S   13W   34   NE
     43-11-34N   124-13-18W
     STONE
MONTAIN STREET SITE   C
     25S   13W   09
     43-24-40N   124-14-51W
     SAND & GRAVEL(SAND)
NEW PEART
     27S   13W   36
     43-11-40N   124-11-57W
     COAL

NEWCASTLE
26S   12W   21   "SW,NW"
43-18-09N   124-08-14
COAL   SANDSTONE AND SHALE

NICOLI
33S   12W   23
42-42-29N   124-05-43W
AU   GABBRO

NOBLE CREEK NO 1
27S   13W   12
43-14-30N   124-11-05W

NOBLE CREEK NO 2
27S   13W   12
43-14-43N   124-11-04W

NORTON GULCH CHROMIUM PROSPECT
27S   14W   09   SW
43-19-33N   124-22-26W
BLACK SANDS

NUGGET
32S   12W   33
42-44-49N   124-06-15W
AU

OLD DIAMOND PIT
31S   12W   13
42-53-12N   124-03-53W   STONE

OLD GOLD
32S   12W   29
42-45-30N   124-07-31W
AU

OLD PEART
27S   13W   35
43-11-24N   124-12-30W   COAL

OLDLANDS
26S   13W   08
43-19-38N   124-16-30W
COAL

OVERLAND
27S   13W   09   NE
43-14-50N   124-14-32W
COAL   SANDSTONE AND SHALE

PANTER MINE
28S   13W   17   SW
43-08-31N   124-16-37W
COAL

PARKER MAGNET PROCESS
25S   13W   34
43-21-52N   124-13-21W

PIGEON POINT

```
        25S     13W     30
        43-22-10N       124-17-38W

PIONEER MINE
        27S     14W     33      "W,W,NE"
        43-11-37N       124-21-47W
        CR      BLACK SANDS IN BEACH SANDS


PROSPECT
        27S     13W     03      "SW,SW"
        43-15-21N       124-14-17W
        AU
PURKENSON QUARRY AND PLANT
        28S     14W     30
        43-07-18N       124-24-21W                      STONE
REDBIRD & SCORPION
        33S     12W     28
        42-41-52N       124-07-57W
RESEVOIR
        25S     13W     27      "S,SW"
        43-22-08N       124-14-05W
        COAL    SANDSTONE AND SHALE
REYNO CLAIM
        32S     12W     30
        42-45-33N       124-08-55W
        SILTSTONE
RHODA CREEK PIT
        29S     12W     33      "NE,SW"
        43-00-40N       124-07-54W
        SAND & GRAVEL           GRAVEL
RIVERTON COALS
        28S     13W     17      "SW,SW"
        43-09-02N       124-16-31W
        COAL    SANDSTONE AND SHALE
ROBERTS
        33S     12W     26
        42-41-49N       124-06-03W


ROBERTSON PIT & PLANT
        28S     14W     30
        43-07-05N       124-24-35W
        SAND & GRAVEL
ROCK CREEK NO 1
        33S     12W     33
        42-40-48N       124-08-06W
        CR
ROCKY BAR
        32S     11W     22
```

42-46-38N   123-58-20W   AU

RODGERS SITE C
      23S   13W   13   SW
      43-34-17N   124-11-42W
      SAND & GRAVEL(SAND)
ROOKARD COAL PROSPECT
      29S   10W   06
      43-04-30N   123-55-14W
ROOKARD MANAGANESE PROSPECT
      29S   11W   33
      43-01-15N   124-00-18W
ROSE MINE
      27S   14W   21   NE
      43-13-14N   124-21-47W
      MASSIVE BLACK SANDS

ROSEBURG LUMBER CO
      25S   13W   01
      43-25-48N   124-15-28W
      SAND & GRAVEL
SACCHI BEACH CHROMIUM OCCURRENCE   26S   14W
      32   S
      43-16-14N   124-22-41W
      BLACK SANDS
SALMON MT.
      32S   12W   19
      42-46-46N   124-09-01W
      AU   BASALTS & SEDIMENTS

SANDSTONE QUARRY C
      28S   10W   26   SW
      43-06-00N   123-51-20W
      STONE(SANDSTONE)
SCORBY & MCGINITY
      28S   13W   05
      43-10-08N   124-15-52W   COAL
SECTION 10 CHROMIUM OCCURENCE   27S   14W
      10   SW
      43-14-16N   124-20-42W
      BLACK SANDS
SECTION 26 PIT
      29S   12W   26   "NW,NW"
      43-02-04N   124-05-46W
      SAND & GRAVEL   GRAVEL
SECTION 27 CHROMIUM OCCURENCE
      27S   14W   27   SE
      43-11-55N   124-20-24W

BLACK SANDS
SECTION 3 CHROMIUM OCCURENCE
        27S   14W   03
        43-15-09N   124-21-20W
        BLACK SANDS

SECTION 32 CHROMIUM OCCURENCE
        28S   14W   32   "N,SE"
        43-06-18N   124-22-57W
        BLACK SANDS
SECTION 32 PIT
        29S   14W   32   "NW,NE"
        43-01-30N   124-22-51W
        SAND & GRAVEL   GRAVEL
SECTION 33 CHROMIUM OCCURENCE
        28S   14W   33   N
        43-06-22N   124-22-04W
        BLACK SANDS
SECTION 36 CHROMIUM OCCURENCE
        26S   14W   36   "N,SW"
        43-16-43N   124-18-03W   BLACK SANDS
SECTION 5 CHROMIUM OCCURRENCE
        27S   14W   05   NE
        43-15-22N   124-23-07W
        BLACK SANDS
SELL
        28S   13W   20
        43-07-41N   124-16-05W

SENGSTACKEN
        26S   14W   25   "SE,NE"
        43-17-21N   124-17-59W
        BLACK SANDS IN FINE IRON
SEVEN DEVILS
        26S   14W   32
        43-16-20N   124-22-55W
SEVEN DEVILS MINE
        27S   14W   10
        43-14-34N   124-20-49W
SEVENMILE CREEK NO 1
        27S   14W   23
        43-12-49N   124-19-36W
SEVENMILE CREEK NO 2
        29S   14W   10
        43-04-27N   124-20-55W
SHEPARD CHROMIUM OCCURRENCE
        27S   14W   16   SW
        43-13-47N   124-21-50W
        MARINE BEACH AND OFFSHORE DEPOSITS
159

SHILLER
    32S   12W   28
    42-46-07N   124-07-07W   AU

SMITH
    25S   12W   27
    43-22-23N   124-06-20W   STONE
SMITH POWER COAL PROSPECT
    27S   13W   01   "NE,NE"
    43-15-57N   124-10-49W
SMITH-KAY
    28S   13W   17   "S,NW"
    43-08-44N   124-16-26W
    COAL   SANDSTONE AND SHALE
SMITH-POWERS
    26S   13W   36
    43-16-01N   124-10-52W   COAL
SNOWBIRD
    33S   12W   15
    42-43-18N   124-07-06W
SOUTH SLOUGH
    27S   14W   02   S
    43-15-09N   124-19-29W
    COAL   SANDSTONE AND SHALE
SOUTHPORT
    26S   13W   22   "SE,SE,SE"
    43-18-12N   124-13-12W
    COAL   SANDSTONE
STAINBECK
    27S   12W   31
    43-11-24N   124-10-11W   COAL
STANDLEY
    27S   13W   13
    43-14-02N   124-11-43W
STATSMAN
    30S   14W   18
    42-58-12N   124-24-05W
STRIPPING COAL PROJECT
    28S   13W   08
    43-09-09N   124-15-52W   COAL
SUGARLOAF MOUNTAIN PIT
    29S   12W   12   SW
    43-04-13N   124-04-33W
    SAND & GRAVEL   GRAVEL
SUMMER NO 1
    26S   12W   32
    43-16-30N   124-09-00W
SUMMER NO 2
    26S   12W   29

160

```
          43-17-08N          124-09-26W
SUMMER NO 3
          27S       13W      12
          43-14-44N          124-10-46W
SUMMER NO 4
          26S       12W      17
          43-19-10N          124-09-30W

SUMMER NO 6
          26S       12W      31
          43-16-25N          124-10-00W              COAL
SUMMER NO 9
          26S       12W      28
          43-17-21N          124-08-17W
SUNSET BAY
          26S       14W      09
          43-20-03N          124-22-17W
          SAND & GRAVEL
THE LAGOONS
          27S       14W      32
          43-11-04N          124-23-04W
          BLACK SANDS
THIRTY SIX
          26S       13W      25      SW
          43-17-12N          124-11-40W
          COAL    SANDSTONE AND SHALE

THOMAS
          26S       13W      22      SE
          43-17-52N          124-13-17W
          COAL    SANDSTONE AND SHALE
TWOMILE CREEK CHROMIUM OCCURENCE
          27S       14W      15      CENTER
          43-13-49N          124-20-45W
          BROWN AND BLACK STAINED SAND
TWOMILE CREEK CHROMIUM OCCURENCE
          29S       15W      11
          43-03-23N          124-26-05W              BLACK SANDS
US MINING CO
          27S       14W      16      NW
          43-13-59N          124-21-57W
          MARINE:  BEACH AND OFFSHORE SEDIMENTS
VAUGHAN
          26S       12W      20
          43-18-11N          124-08-57W
          COAL
VEY
          26S       13W      05
          43-20-18N          124-16-32W
```

WARD
       25S   12W   19
       43-23-11N   124-09-42W

WARNER CREEK PIT
       29S   12W   33   "N,NW,NE"
       43-02-22N   124-07-38W
       SAND & GRAVEL   GRAVEL
WEST
       26S   12W   31
       43-16-24N   124-10-12W   COAL
"WHITE BRIDGE, POWERS HIGHWAY"   31S   12W
       17
       42-53-12N   124-08-42W
       SAND & GRAVEL
WHITE ROCK
       32S   12W   21   "SE,SE"
       42-47-01N   124-06-35W
       CR   SERPENTINE
WILEY
       29S   12W   28   "E,CENTER"
       43-01-41N   124-07-30W
       SAND & GRAVEL

WISKEY RUN
       27S   14W   20
       43-12-41N   124-23-36W
       AU
WOOMER
       27S   14W   36
       43-10-57N   124-18-18W
WORTH
       25S   12W   29
       43-22-38N   124-09-17W

WTILLANCH SLOUGH NO 1
       25S   12W   07
       43-24-54N   124-09-48W

WTILLANCH SLOUGH NO 2
       25S   12W   18
       43-24-09N   124-09-32W

YOKAM POINT

       26S   14W   04

       43-20-36N   124-21-33W

CURRY County

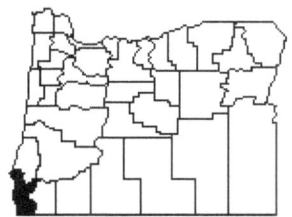

NAME
TOWNSHIP RANGE SECTION   SECTION FRACTION
NORTH LATITUDE   WEST LONGITUDE
MINED MATERIAL

Last Update of Claim or Quarry if I was able to get it

ADOLPHSON & NODINE
>31S 15W 24
>42-52-47N 124-23-16W

AGNESS GROUP
>35S 11W 30
>42-30-58N 124-03-41W
>SHEARED GREEN TALCOSE SERPENTINE
>81 02

AGNESS PASS PIT
>35S 11W 07
>42-33-49N 124-03-24W
>STONE

AGNESS RD
>36S 14W 11
>42-28-32N 124-19-55W
>SAND & GRAVEL

ALEXANDER BROS.
>32S 10W 31
>42-44-52N 123-54-19W
>GREENSTONE (#PORPHYRY#) 81 02

AMAMS PROSPECT
>37S 12W 01
>42-24-13N 124-03-36W

ANDERSON SITE
>33S 15W 10 SE
>42-43-59N 124-27-22W
>STONE

ARCH ROCK
>32S 13W 21
>42-46-48N 124-12-25W

ARNTZENS
>36S 14W 30
>42-25-59N 124-24-19W
>SAND & GRAVEL

B & C LATERITE
>38S 10W 11
>42-16-20N 123-51-04W

B & K NO.1
>37S 12W 20
>42-21-55N 124-08-46W
>78 11

BABYFOOT
>38S 09W 30
>42-14-14N 123-49-01W
>CR SERPENTINE 81 01

BABYFOOT MERCURY
    38S   09W   31   NW
    42-13-44N   123-48-41W
    GREENSTONE TUFFS AND LAVAS
BAILEY CABIN PROSPECT
    39S   10W   03
    42-12-15N   123-52-10W
    GREENSTONE AND SERPENTINE
    81 04
BAILEY CHROMITE
    39S   10W   03
    42-12-11N   123-51-18W
    CR   SERPENTINE
    81 02
BAILEY CHROMITE PROSPECT
    39S   10W   03   SE
    42-12-04N   123-51-19   CR
BALDFACE PLACERS
    40S   10W   20
    42-04-03N   123-53-40W
    GREENSTONE AND METAGABBRO

BALDFACE RIDGE
    40S   10W   30
    42-04-22N   123-54-07W
    SERPENTINE   81 02
BALDFACE RIDGE LATERITES
    40S   11W   25
    42-03-17N   123-56-21W
BANKUS BAR
    38S   14W   29
    42-15-51N   124-22-47W
    SAND & GRAVEL
BASALT QUARRY
    35S   14W   04   SE
    42-34-13N   124-21-37W
    STONE(BASALT)

BASALT QUARRY
    35S   14W   10   NW
    42-34-00N   124-20-45W
    STONE(BASALT)
BEAR CAT
    2S   13W   22
    42-46-43N   124-13-34W
    SLATE 81 02

BERRY PROSPECT
    36S   11W   31

                42-25-27N   124-02-20W
BIG BEN
                32S   13W   23
                42-46-16N   124-12-14W
                AU AG ARGILLITE
                81 02
BIG CAT
                34S   12W   09
                42-39-17N   124-08-20W
                SERPENTINE   81 02
BIG SUNSHINE
                33S   14W   13   E
                42-42-60N   124-18-11W
                AU PT   STREAM GRAVELS   78 11
BLACK BEAR
                41S   11W   14   NE
                42-00-16N   123-57-16W
                CHERT 81 02
BLACK CUB NO. 1
                41S   11W   01
                42-02-05N   123-56-41W
                SERPENTINE   81 02
BLACK CUB NO. 2
                41S   11W   01
                42-02-12N   123-56-49W
BLACK ROCK # 31
                35S   12W   35
                42-30-18N   124-05-41W
                SERPENTINE   81 01
BLACK ROCK NO. 10
                35S   12W   05
                42-34-46N   124-08-59W
                CR
BLACK ROCK NO. 10
                36S   12W   05
                42-29-25N   124-09-16W
                SERPENTINE   81 02
BLACKBIRD
                33S   12W   35
                42-40-25N   124-05-48W
                CR
BLUE RIBBON #1
                39S   10W   10
                42-11-20N   123-51-55W
                SERPENTINIZED PERIODOTITE

BLUE RIBBON #3
                39S   10W   10
                42-11-20N   123-52-00W

SERPENTINIZED DUNITE

BONANZA
    33S   12W   04
    42-39-41N   124-08-47W
    AU   SERPENTINE   81 04
BONANZA BASIN - BOULDER CREEK PLACERS
34S   12W   08   "NW,NW"
    42-39-17N   124-09-48W
    AU CR   STREAM AND BENCH GRAVELS.
    81 04
BOULDER CREEK MINING CO STAR MINING   34S   13W
    25
    42-36-19N   124-11-33W
    AU
BOWSER
    38S   10W   23   W
    42-15-18N   123-51-13W
    CR   SERPENTINE   81 02
BUCK CHROMITE PROSPECT
    39S   10W   11
    42-11-22N   123-50-57W
    SERPENTINE   81 02
BUCKSKIN RIDGE
    40S   10W   12   SE
    42-05-52N   123-49-53W
    LATERITES
BUFFINGTON PIT
    35S   14W   09   NW
    42-33-59N   124-22-23W
    STONE(BASALT)
"BULLARDS SAND AND GRAVEL, INC."
    31S   15W   12
    42-53-56N   124-25-04W
    SAND & GRAVEL
BUNKER HILL GROUP
    37S   12W   20
    42-22-09N   124-08-45W
    SERPENTINE   81 02
BURNED CABIN CLAIM
    39S   10W   02
    42-12-13N   123-50-37W
    CR   SERPENTINE   81 02

BUTCHER HILL QUARTZ
    32S   13W   21
    42-46-27N   124-14-13W
    METAVOLCANICS AND METASEDIMENTS

CAMPBELL PIT
>41S 13W 03 NE
>42-03-31N 124-13-22W
>STONE

CAPE BLANCO
>32S 16W 12
>42-48-15N 124-31-43W
>AU PT OS CR BLACK SANDS 81 04

"CARPENTER, CHANCEY"
>36S 14W 30
>42-25-58N 124-24-18W
>SAND & GRAVEL

CARTER CREEK DIVIDE
>39S 10W 02
>42-12-29N 123-50-37W
>CR SERPENTINE 81 02

CEDAR SPRINGS LATERITE
>40S 10W 35
>42-02-40N 123-51-06W
>SERPENTINE 81 02

CHETCO CLAIM
>38S 12W 10
>42-16-34N 124-05-36W
>SERPENTINE

CHETCO COPPER CO
>39S 10W 12
>42-11-16N 123-49-07W

CHETCO LAKE OCCURRENCE
>39S 11W 23
>42-09-42N 123-57-42W
>DUNITE 81 02

CHETCO LAKE RHODONIT
>39S 11W 25 SW
>42-08-43N 123-56-56W
>QUARTZITE AMPHIBOLITE 81 02

CHROME BEACH PLACER
>31S 15W 09
>42-53-55N 124-29-17W
>BEACH SAND

CHROME CREEK LATERITE
>40S 10W 07
>42-06-50N 123-54-20W
>SERPENTINE 81 02

CHROME OCCURRENCE
>37S 12W 16 SE
>42-22-11N 124-06-27W   81 04

CHROME OCCURRENCE
>37S 12W 16

```
              42-21-47N       124-07-33W
              SERPENTINE   81 03
CHROME OCCURRENCE
              39S      10W      20
              42-09-55N       123-53-45W
              UHRAMAFICS
CHROME PROSPECT
              38S      10W      36        "W,NW"
              42-13-26N       123-49-59W                          81 03
CHROME PROSPECT
              39S      10W      26        NW
              42-09-01N       123-50-50W
              ULTRAMAFICS
CHROME PROSPECT  NO. 73
              39S      10W      26        SW
              42-08-35N       123-51-00W
              ULTRAMAFICS
CHROMITE FLOAT
              39S      10W      03
              42-12-35N       123-52-05W
              SERPENTINE   81 02
CHROMITE FLOAT OCCURRENCE
              39S      10W      03        NW
              42-12-44N       123-52-09W
              MN       CHERT 81 01
CHROMITE OCCURRENCE
              39S      10W      08        CENTER
              42-11-29N       123-54-08W
              ULTRAMAFICS 81 03
CHROMIUM PROSPECT
              38S      10W      14
              42-15-42N       123-50-55W
CHROMIUM PROSPECT
              38S      10W      14
              42-15-28N       123-51-10W
CHROMIUM PROSPECT
              38S      10W      36
              42-13-31N       123-49-49W
CLAPSHAW
              31S      15W      35
              42-50-50N       124-27-07W
              CHERT?        81 02

CLARNO
              35S      14W      09        CENTER
              42-33-38N       124-21-12W

CLAY HILL
              34S      11W      02
```

42-40-12N   123-58-37W

CLEOPATRA-TAYLOR CREEK
      40S   10W   32
      42-02-02N   123-54-52W
      SERPENTINE   81 02
CLIFFSIDE
      33S   14W   17
      42-43-18N   124-22-46W
      DIORITE   81 02
COBALT GROUP PROSPECTS
      36S   11W   33
      42-25-09N   124-00-23W
      SERPENTINE   81 02
COBALT PROSPECT
      39S   10W   26   "SW,SW,NW"
      42-08-50N   123-51-14W
      PYROXENITE   81 03

COLBROOK PIT
      34S   14W   30   SE
      42-35-58N   124-23-49W
      STONE
COLDIRON PIT
      37S   14W   22
      42-21-07N   124-20-46W
      STONE
COLEGROVE
      40S   14W   02
      42-08-44N   124-19-41W
      CHERT
COLLIER CREEK
      36S   12W   36
      42-24-53N   124-08-45W
COLLIER CREEK
      37S   12W   01
      42-23-29N   124-02-57W
      LATERITES
COLLIER CREEK COPPER CO
      36S   11W   30
      42-25-58N   124-02-54W
      CU
COLLIER CREEK COPPER NO. 13
      37S   12W   12
      42-23-17N   124-03-42W
      SERPENTINE   81 02
COLLIER CREEK CRAGS
      37S   12W   08
      42-23-14N   124-08-21W

SERPENTINE   81 02

COLLIER CREEK LATERITES
      37S   12W   01
      42-24-07N   124-03-23W
COLLINS
      36S   15W   01
      42-29-08N   124-24-00W
      AU   BLACK SANDS 78 11

COMBINATION
      32S   13W   22
      42-46-25N   124-13-34W
      AU AG ARGILLITE   81 02
COME & GET IT CLAIM
      40S   11W   36
      42-02-22N   123-56-48W
      DUNITE
CONN CREEK QUARRY
      37S   14W   21   NW
      42-21-45N   124-22-19W
      STONE
COPPER CANYON PROSPECT
      35S   12W   11   SW
      42-33-29N   124-06-13W
      SUBMARINE VOLCANICS
COPPER CITY PROSPECTS
      37S   12W   01
      42-24-09N   124-02-58W
      CU   SERPENTINE   81 02

CORBIN
      32S   14W   08
      42-48-34N   124-23-55W
      AU PT STREAM GRAVEL   78 11
COTTONWOOD CAMP LATERITE
      39S   11W   22
      42-08-41N   123-58-39W
      LATERITES   81 02

CROCKETT
      38S   14W   21
      42-16-35N   124-21-07W
      SAND & GRAVEL
CROOKED RIFFLE BHR & PLANT
      35S   12W   11
      42-33-36N   124-05-25W
      SAND & GRAVEL

CRYSTAL CREEK
      32S   15W   01
      42-49-58N   124-25-10W

CRYSTAL CREEK MANGANESE
      31S   15W   26
      42-51-54N   124-26-23W
      CHERT
CRYSTAL TERRACE PLACER
      31S   15W   35
      42-50-55N   124-27-01W
      BEACH SANDS
DIAMOND CREEK PLACER
      41S   10W   16
      42-00-26N   123-52-43W
      HG   PROPYLITIZED DIORITE   81 01
DIAMOND FLAT
      41S   10W   02
      42-01-25N   123-52-11W
      SERPENTINIZED HARZBURGITE WITH A FEW
      PATCHES OF DU   81 03
DIAMOND NICKEL
      41S   10W   10
      42-01-04N   123-52-25W

DIAMOND PROSPECT
      41S   10W   11   S
      42-00-37N   123-50-43W
      MAFIC TO INTERMEDIATE INTRUSIVE ROCKS
DINAWADJA
      32S   10W   33   SW
      42-44-33N   123-53-13W
      DACITE   81 03

DIVELBISS #2
      32S   14W   08
      42-48-27N   124-23-30W
      AU PT   STREAM GRAVEL   78 11
DOE GAP
      39S   10W   34
      42-07-49N   123-52-13W
      LATERITES   81 02
E.L.BENNETT ESTATE
      40S   13W   34
      42-04-13N   124-13-51W
      SAND & GRAVEL
EAGLE CREEK GROUP

38S   10W   24   NW
42-15-09N   123-49-55W
"METAVOLCANICS, METASEDIMENTS"   81 02

EAGLE MOUNTAIN GOSSAN
        38S   10W   13   NE
        42-16-05N   123-49-24W
        METAVOLCANIC   81 02
EAGLE MOUNTAIN PROSPECT
        38S   10W   13   "NE,NE,NW"
        42-16-16N   123-49-33W
        SERPENTINITE 81 02
EAGLE'S NEST CLAIM
        38S   10W   11
        42-17-04N   123-50-26W
        PERIDOTITE   81 02
EAST FORK PANTHER PIT
        36S   14W   31
        42-24-57N   124-24-45W
        STONE
ECKLEY PROSPECT
        31S   13W   35
        42-50-13N   124-12-19W
        SANDSTONE AND SHALE
EDNA FRY
        34S   12W   14
        42-38-03N   124-05-49W
        SERPENTINE   81 02
ELK RIVER PLACERS
        33S   14W   23
        42-42-28N   124-19-51W
        AU PT   GRAVELS
ELMER BANKUS
        40S   13W   33
        42-04-05N   124-14-37W
        SAND & GRAVEL
ELMER BANKUS QUARRY
        40S   13W   24   NW
        42-06-14N   124-11-45W   STONE

EMILY CABIN
        39S   10W   10
        42-11-14N   123-51-41W
        SERPENTINE   81 02
EMLLY MINE
        39S   10W   10   S
        42-11-15N   123-51-40W
        AU   GREENSTONE (PYROXENE TUFF - JPTA)
        81 02

173

EMPIRE PROSPECT
        38S   10W   12
        42-16-25N   123-49-49W

"FERRY CREEK BAR, PIT, & PLANT"
        40S   13W   32
        42-03-55N   124-15-56W
        SAND & GRAVEL
FLOAT CHROMITE CHROMITE
        36S   14W   08
        42-28-17N   124-22-53W
        SERPENTINE
FLORENCE
        40S   12W   08
        42-06-27N   124-09-03W
        HORNFELSED SHALE   81 02
FLOYD D. SMITH
        35S   14W   08
        42-33-37N   124-23-02W
        SAND & GRAVEL
FOSTER CREEK CLAIM
        34S   12W   12
        42-39-14N   124-05-10W
        CR   SERPENTINE   81 02
FRANCINE CLAIM
        39S   10W   11
        42-11-03N   123-51-11W
        SERPENTINIZED DUNITE   81 02
FRANKFORT QUARRY
        36S   14W   31
        42-24-57N   124-24-45W
        STONE
FRAZIER
        38S   10W   26   "E,NW"
        42-14-45N   123-50-44W
        AU AG GREENSTONE 81 02

FREEMAN BAR AND PLANT
        40S   13W   33
        42-04-11N   124-15-07W
        SAND & GRAVEL
GALENA KING
        34S   11W   06
        42-39-19N   124-03-36W
GAME LAKE GROUP
        36S   12W   23
        42-25-59N   124-05-04W
        "PERIDOTITE PORPHYRY, SERPENTINE"   81 02
GARDNER MINE

39S   11W   10
42-11-44N   123-59-03W
CR   SERPENTINE   81 02

GLADE CREEK QUICKSILVER
37S   13W   29
42-20-21N   124-16-11W
SERPENTINIZED ULTRAMAFIC ROCKS   81 03
GOLAY MAGNETITE PLACER
35S   11W   34
42-30-05N   123-59-49W
GOLD BAR
34S   11W   19
42-37-16N   124-03-50W
AU
GOLD BASIN
37S   10W   33   SW
42-18-24N   123-53-22W
GRAVELS

GOLD BASIN PLACERS
37S   10W   33
42-18-07N   123-53-19W
GOLD BEACH CRUSHING OPERATION
32S   13W   09
42-48-18N   124-14-39W
SAND & GRAVEL
GOLD MINE
35S   14W   18   SE
42-32-33N   124-23-50W
AU
GOLD MINE
40S   12W   28   "SW,NW"
42-03-51N   124-08-01W
AU
GOLD OCCURRENCE
37S   12W   03
42-23-56N   124-05-58W
SANDSTONE   81 03
GOLD PLACER
38S   10W   15
42-15-43N   123-52-10W
GOLDEN DREAM
38S   10W   12   W
42-16-42N   123-50-06W
AU AG GREENSTONE
SLATE SERPENTINE   81 02
GOLDEN EAGLE

38S   10W   24   "SE,NE"
42-15-08N   123-49-01W
AG   SERPENTINE AND ARGILLITE

GOLDEN ECONOMY
32S   10W   31   "SW,SW"
42-44-28N   123-54-19W
META-ANDESITE PORPHYRY   81 02
GOLDEN FRACTION
32S   10W   31   "N,NW"
42-44-44N   123-54-24W
GREENSTONE PORPHYRY   81 02
GOLDEN OAK
32S   10W   31   "NE,NE"
42-45-02N   123-54-37W
GREENSTONE 81 02
GOLDEN RATTLER GROUP
32S   10W   32   SW
42-44-30N   123-54-10W
META-ANDESITE PORPHYRY   81 02
GOOD LUCK
32S   10W   32   SE
42-44-30N   123-54-12W
META-ANDESITE PORPHYRY DACITE 81 02
GRAY BUTTE
36S   12W   12
42-28-27N   124-03-19W
LATERITES   81 04
GRAY COPPER NO. 18
36S   12W   36   S
42-24-38N   124-03-44W
SERPENTINE

HALF MOON BAR
33S   11W   25
42-41-47N   123-56-15W
AU

HAMAKER GROUPS
38S   09W   31
42-13-35N   123-48-28W
HANSCUM
38S   10W   14
42-16-21N   123-50-36W
CR   SERPENTINIZED DUNITE   81 02
HARDEN BROOK MANGANESE
38S   14W   34

42-14-37N   124-20-43W
CHERT AND SERPENTINE

HARMONY PROSPECT
32S   13W   35
42-45-13N   124-12-24W
CONGLOMERATE SANDSTONE SHALE
81 03

HARRY B.
38S   10W   11
42-17-07N   123-50-23W
PARTLY SERPENTINIZED DUNITE   81 02

HAWKS REST VIEW
39S   10W   10
42-11-29N   123-52-09W
CR   DUNITE   81 02

HAY PROSPECT
38S   10W   11
42-16-46N   123-50-45W

HILLTOP MINE GROUP
38S   10W   36
42-13-15N   123-48-56W
JUR   81 03

HILLTOP PROSPECT
38S   09W   31   SW
42-13-05N   123-48-38W
SILICEOUS METASEDIMENTS 81 03

HORSEHEAD
36S   11W   17
42-27-41N   124-01-31W
SANDSTONE   81 0

HUBBARD CREEK BEACH PLACER
33S   15W   09
42-44-14N   124-28-41W
BEACH SANDS

HUBBARD MOUND BEACH
36S   15W   12   E
42-28-34N   124-25-18W
BLACK SANDS

HUSTIS
38S   10W   14   CENTER
42-15-57N   123-50-46W
AU AG "METAVOLCANICS, VOLCANIC WACKE"

ILLAHE GROUP
34S   12W   13
42-38-11N   124-04-41W
SHEARED GREEN SERPENTINE OR PERIDOTITE-
PORPHYRY   81 03

INDEPENDENCE NO. 1 & 2
39S   10W   14

42-10-30N 123-50-09W MICA
INDIGO CREEK
     36S 11W 04
     42-29-26N 124-00-27W
     CR SERPENTINE WITH SANDSTONE AND
     GABBRO TO THE EAST 81 03

IRENE CHROMITE
     41S 11W 15 CENTER
     42-00-36N 123-58-01W
     CR SERPENTINE 81 03
IRON MOUNTAIN
     33S 12W 33
     42-40-04N 124-08-20W
     LATERITES 81 04
IRON MT QUARRY
     33S 13W 25
     42-41-29N 124-12-02W
     STONE
IRON PROSPECT
     38S 10W 31 SE
     42-12-59N 123-55-15W
     "PYROXENITE, GABBRO" 81 03
J C COMPTON CO
     32S 15W 35
     42-45-54N 124-26-44W
     SAND & GRAVEL
JOE BLANCHARD
     32S 15W 11
     2-49-03N 124-26-27W
     STONE

JOE HALL PROPERTY
     40S 13W 33 NW
     42-04-21N 124-15-26W
     STONE
KALAMAZOO BEACH PLACER
     34S 14W 20
     42-36-55N 124-23-24W
     AU CR BEACH SANDS
KESSLER & FRY'S
     36S 12W 22
     42-26-37N 124-05-24W GREENSTONE
SERPENTINE SCHIST 81 03

KEYSTONE
     33S 10W 17
     42-41-47N 123-54-45W

AU AG GREENSTONE METAGABBRO METAGABBRO
81 03
KNAPP PIT
32S   15W   20
42-47-08N   124-30-23W
SAND & GRAVEL

LANCASTER
32S   10W   30   SE
42-45-14N   123-54-29W
ANDESITE PORPHYRY PORPHYRY   81 03
LAWRENCE BRADY PROSPECT
34S   12W   09
42-39-07N   124-08-28W
LAWRENCE MANGANESE
39S   14W   10
42-12-38N   124-20-28W
CHERT AND SERPENTINE

LEITH MANGANESE
36S   14W   31
42-24-53N   124-24-31W
LENA CLAIM
37S   12W   29
42-21-09N   124-08-46W
SERPENTINE   81 03
LITTLE BOY NO. 1
39S   10W   02
42-12-25N   123-50-58W
CR   SERPENTINE   81 03
LITTLE BOY NO. 4
39S   10W   02
42-12-22N   123-50-34W
LITTLE BOY NO. 5
39S   10W   02
42-12-35N   123-50-36W
LITTLE SIBERIA
38S   10W   11
42-16-33N   123-50-48W
CR   DUNITE   81 03
LLOYD MANGANESE
36S   14W   29
42-25-54N   124-22-33W
CHERT
LONE RANCH PRICEITE
40S   14W   22   NE
42-06-07N   124-20-38W

BORON SERPENTINE   81 04
LONG RIDGE
  38S 12W 13
  42-15-58N 124-04-10W
  MN CHERT 81 03
LOST IS FOUND
  38S 10W 11
  42-16-37N 123-50-15W
  SERPENTINE 81 03

LOST LEE CLAIM
  41S 10W 02 SE
  42-01-40N 123-50-22WCR "HARD, BLOCKY
  SERPENTINE ALTERED FROM OLIVINE-RICH"
  81 03

LOWER LARSON
  36S 12W 10
  42-27-14N 124-05-38W
  LATERITES 81 04
LUCKY DAY PROSPECT
  38S 09W 30
  42-14-07N 123-48-21W
  SERPENTINE 81 03
LUCKY NO. 1
  41S 10W 16
  42-59-59N 123-53-12W
LUCKY WARREN
  40S 12W 07
  42-05-55N 124-09-00W
  SHALE 81 03
LUDLUM PIT
  41S 12W 03
  42-01-58N 124-06-32W
  STONE
M C MINE
  39S 10W 01 CENTER
  42-12-29N 123-49-25W
  METAVOLCANICS AND METASEDIMENTS 81 04
MADDEN
  32S 15W 04
  42-49-57N 124-28-23W
  AU PT ANCIENT BEACH GRAVEL
MAMMOTH
  33S 10W 03 NW
  42-44-09N 123-52-22W
  AU AG METAGABBRO 81 03
MANGANESE PROSPECT
  35S 12W 16

42-32-48N   124-08-15W

MANGANESE PROSPECT
     39S   10W   15
     42-10-09N   123-51-30W
     TUFFACEOUS METASEDIMENTS

MARCELLA
     38S   10W   12
     42-16-53N   123-50-11W
     PARTLY SERPENTINIZED DUNITE

MARIGOLD
     32S   10W   33
     42-44-34N   123-53-02W   AU AG

MC CALEB MINE
     38S   10W   11
     42-16-42N   123-50-06W
     CR   SAXONITE   81 03

MCADAMS
     30S   14W   20
     42-56-58N   124-23-04W   MN

MCCALEB'S SOURDOUGH NO. 1
     38S   10W   12   "W,NW"
     42-16-55N   123-50-04W
     WEATHERED TAN PERIDOTITE WITH TALC
     CRYSTALS

MCCORMICK MINING & MINERAL CO
     32S   14W   08
     42-48-25N   124-23-08W   AU

MCKENZIE SITE
     32S   15W   08   SW
     42-48-41N   124-30-32W
     SAND & GRAVEL   93 04

"MCKENZIE, THOMAS M"
     40S   13W   33
     42-04-15N   124-15-02W
     SAND & GRAVEL

MEADOW CREEK PIT & PLANT
     36S   14W   31
     42-24-57N   124-24-45W   STONE

MEEKS PLACER
     32S   15W   33
     42-45-39N   124-28-30W
     AU PT   BEACH SANDS

MELANTERITE TUNNEL
     39S   10W   10   NW

```
              42-11-33N        123-52-10W
              CU       TUFFACEOUS METASEDIMENTS
MIDDLE FORK SIXES RIVER
              32S       13W      14
              42-46-46N        124-12-10W
              SANDSTONE AND SHALE
MILLER CREEK GOSSAN
              38S       10W      13       "W,SW"
              42-15-36N        123-49-58W
              "ALTERED TUFFS, FLOWS"       81 03
MINDORO PROJ.
              36S       12.5W   24
              42-26-31N        124-10-12W
              AU AG CU PB ZN PT NI CO CR          93 04
MISLATNAH PROSPECT
              37.5S     12W      34
              42-18-24N        124-06-06W                        81 04
MISS DOLLY
              34S       12W      05       NE
              42-40-01N        124-08-10W
              SERPENTINE
MORNING SUN PROSPECT
              39S       10W      10
              42-11-16N        123-51-53W
              CR        81 03
MORRISON GULCH
              38S       10W      23       SW
              42-14-38N        123-50-54W
              SERPENTINE   81 03
MOSS ROSE
              33S       14W      16       SW
              42-42-47N        124-22-13W              GREENSTONE
              81 03
MULE CREEK AREA PLACERS
              33S       10W      10
              42-43-18N        123-52-37W
              AU        CREEK GRAVELS       81 04
MULE MTN.
              33S       10W      17
              42-42-07N        123-54-13W
              AU AG  GREENSTONE  81 03
MYERS CREEK QUARRY
              38S       14W      07
              42-18-14N        124-24-16W
              STONE
NANCY HANK CLAIM
              39S       11W      10
              42-11-21N        123-59-21W     81 03
NEW DISCOVERY
                            182
```

```
            37S      14W      06
            42-24-19N         124-24-17W


NEWHOUSE PROSPECT
            31S      15W      23       SW
            42-57-11N         124-27-14W
            CHERT AND SANDSTONE
NIGHTHAWK PROSPECT
            35S      11W      33
            42-30-20N         124-00-59W
            MAFIC GREENSTONE
NORTH FORK GRAVEL PIT
            31S      13W      14       "SW,SE"
            42-52-35N         124-12-25W
            SAND & GRAVEL         GRAVEL
NORTH STAR
            34S      12W      13
            42-37-52N         124-04-57W
            CR            81 03
NUG AND BOOMER CLAIMS
            37S      12W      29
            42-19-34N         124-08-54W
            ULTRAMAFICS 81 03
OCCURRENCE
            39S      11W      23
            42-07-00N         123-57-36W
OLD RED MINE
            32S      10W      29       S
            42-45-17N         123-53-53W
            METAVOLCANIC         81 03
OLIVE B.
            38S      10W      12
            42-16-33N         123-50-05W
ORIENTAL NO. 1
            39S      10W      10
            42-11-17N         123-52-16W
OTTER POINT BEACH
            36S      15W      13       NE
            42-27-34N         124-25-15W
            BLACK SANDS
PARADISE
            32S      10W      27
            42-45-21N         123-52-01W
            AU       METAVOLCANICS / METAGABBRO
PARKER ELECTROMAGNETIC MACHINE PROJECT       35S
            15W      08
            42-33-45N         124-23-11W
            AU PT             78 11
```

PARKER PLACER
        34S   12W   04
        42-39-46N   124-08-45W   AU

PASTORELLI MINING CLAIM
        37S   12W   29
        42-20-16N   124-09-13W
PEARSOLL GROUP
        38S   10W   02
        42-17-24N   123-50-24W   81 04
PEARSOLL PEAK
        38S   10W   02
        42-17-49N   123-50-46W   CR   81 04
PEARSOLL PEAK DUNITE
38S   10W   11   W
        42-16-48N   123-51-05W
        DUNITE   81 04
PECK MINE
        38S   10W   23
        42-15-12N   123-50-12W
        AU AG GREENSTONE 81 04
PHYLLIS CLAIM
        37S   12W   32
        42-20-30N   124-08-17W
        ULTRAMAFICS 81 04

PINE FLAT
        35S   12W   26
        42-31-22N   124-05-47W
        CU   SERPENTINE   81 04
PINES
        38S   12W   11
        42-16-50N   124-05-25W   81 04
PINNACLE PEAK
        36S   14W   31
        42-24-57N   124-24-45W   STONE
PINNACLE POINT OCCURRENCE
        33S   10W   19   NE
        42-38-31N   123-55-34W

PIONEER PLACER
        31S   15W   15
        42-53-20N   124-27-45W
        BEACH SANDS
PORT OF BROOKINGS
        41S   13W   05
        42-03-07N   124-15-51W
        SAND & GRAVEL

PORT OF GOLD BEACH #2
    36S   15W   36
    42-25-09N   124-25-08W
    SAND & GRAVEL

PORT OF GOLD BEACH #3
    36S   15W   25
    42-25-51N   124-25-09W
    SAND & GRAVEL
PROSPECTORS DREAM
    38S   10W   11
    42-16-38N   123-51-13W   81 04

RED CUB
    34S   11W   02   "SW,SW"
    42-35-05N   123-59-05W
    METAVOLCANICS
RED FLAT LATERITE
    37S   13W   30
    42-20-30N   124-17-23W
    LATERITE AND SAPROLITE DERIVED FROM PARTLY
    SERPENT   81 03
RED FLAT PLACERS
    37S   14W   24
    42-21-28N   124-17-16W   RESIDUAL GRAVELS

RED RIDGE MINING CO.
    37S   13W   18
    42-22-00N   124-17-36W
RED RIVER MINING CO. PLACER
    33S   10W   09   "S,S,CENTER"
    42-42-51N   123-53-12W
    GRAVELS   81 04

RED ROCK PIT
    30S   15W   36   SE
    42-55-42N   124-24-59W
    STONE
RHYOLITE
    32S   10W   30   CENTER
    42-45-37N   123-54-57W
ROBERT E. MINE
    38S   10W   23   "NE,NW"
    42-15-27N   123-50-42W
    AU AG GREENSTONE
ROBERTS PROSPECT
    31S   14W   13   NW
    42-53-01N   124-18-54W
    CHERT
185

ROGUE RIVER BAR
     36S   14W   26
     42-25-48N   124-19-27W
     SAND & GRAVEL
ROGUE RIVER BEACHES
     36S   15W   13
     42-25-00N   124-25-00W
     CR   BLACK SANDS 79 01
ROSIE CLAIM
     39S   11W   11
     42-11-48N   123-57-17W
     CR   81 04
"RUSH, RODNEY K."
     40S   13W   34
     42-03-57N   124-13-25W
     SAND & GRAVEL
RUSSELL F HILL
     35S   11W   06
     42-34-18N   124-03-09W
     SAND & GRAVEL
SAND & GRAVEL PIT
     37S   14W   19   "NE,NE"
     42-21-51N   124-23-50W SAND & GRAVEL

SAND PIT
     38S   14W   30   NE
     42-15-47N   124-23-55W
     SAND & GRAVEL(SAND)   93 04
SAUNDERS PIT AND PLANT
     36S   14W   31
     42-24-57N   124-24-45W
     STONE
SECTION 27 GOLD PROSPECT
     32S   13W   27   "NW,NW"
     42-46-05N   124-13-55W
     METASEDIMENTS   81 04
SHALE PIT
     36S   15W   13
     42-27-52N   124-13-58W
     SHALE

SHASTA COSTA COAL
     35S   11W   05
     42-34-24N   124-02-52W
     COAL

SHASTA COSTA COPPER
> 34S   11W   35   SW
> 42-35-08N   123-59-07W
> PORPHYRITIC GREENSTONE (METABASALT) 81 04

SHYRITE
> 34S   11W   27
> 42-35-54N   124-00-25W
> SERPENTINE

SIGNAL BUTTE
> 36S   14W   36   "NE,NE,NE"
> 42-25-29N   124-18-00W
> CR   81 04

SIXES
> 31S   15W   27   SW
> 42-51-12N   124-27-50W
> AU PT   BEACH GRAVEL   78 11

SIXES RIVER PLACERS
> 32S   14W   11   "NE,SE"
> 42-48-15N   124-19-00W
> AU PT   81 04

SIXES RIVER QUARRY
> 32S   15W   12   NE
> 42-49-19N   124-25-07W   STONE

SLIDE CREEK BAR
> 34S   11W   09
> 42-39-03N   124-00-42W   AU

SMEDBERG BEACH PLACERS
> 37S   15W   01
> 42-23-47N   124-25-22W

SMITH
> 37S   14W   05
> 42-24-21N   124-23-05W   CHERT 81 04

SMITH RIVER
> 41S   11W   10   E
> 42-00-53N   123-58-36W
> LATERITES   81 04

SNOW CAMP
> 37S   12W   19
> 42-20-22N   124-09-14W
> LATERITES   81 04

SNOW CAMP #1
> 37S   12W   28
> 42-19-15N   124-07-50W   ULTRAMAFICS
> 81 04

SOUR DOUGH
> 40S   11W   36
> 42-02-27N   123-56-57W
> CR   DUNITE   81 04

SOURCE C 10
    36S   14W   31
    42-24-57N   124-24-45W   STONE
SOURDOUGH FLAT
    38S   10W   11   SW
    42-16-21N   123-50-36W
LATERITES   81 04

SOURDOUGH NO.1
    38S   10W   12
    42-16-55N   123-50-04W
    PERIDOTITE   81 03
SPOERLS QUARRY
34S   14W   07   "NW,SE"
    42-38-42N   124-23-53W   STONE
SPOKANE CREEK LATERITE
    40S   10W   16
    42-04-23N   123-54-07W
    LATERITES   81 04
SQUARE LAKE LATERITES
    39S   10W   24
    42-09-18N   123-49-56W
SQUARE LAKE PROSPECT
    39S   10W   24   NE
    42-09-57N   123-49-10W   ULTRAMAFICS
SQUIRE
    39S   11W   23
    42-09-48N   123-57-30W   FINE GRAINED
STARR
    36S   13W   31
    42-30-07N   124-17-55W   CU   SERPENTINE
    81 04
STEFFIN MEADO
    36S   14W   31
    42-24-57N   124-24-45W   STONE
STEVENS AND STEAR
    35S   11W   33   "NW,NW"
    42-30-43N   124-01-17W   GREENSTONE
STONE PROSPECT
    38S   10W   12   "NE,SW"
    42-16-36N   123-49-35W
    METAVOLCANICS   81 04
STUMBLE CLAIM
    39S   10W   02   NW
    42-12-37N   123-51-04W
    AU   TUFFACEOUS SEDIMENTS   81 04
SUGAR CREEK GRAVEL PIT
    32S   13W   02   "SE,SE"
    42-49-10N   124-12-00W

SAND & GRAVEL

SUGARLOAF PROSPECT
      38S   10W   25
      42-14-08N   123-49-39W
      SERPENTINE   81 04
SULFIDE OCCURRENCE
      35S   14W   36
      42-30-05N   124-17-26W   METASEDIMENTS
SULLIVAN PIT & PLANT
      31S   15W   12   SE
      42-54-04N   124-25-02W   STONE
SULLIVAN QUARRY
      31S   15W   12   NE
      42-54-26N   124-25-02W   STONE
SUNRISE GOLD MINING CO
      32S   14W   09
      42-48-43N   124-21-29W   AU
SUSIE CLAIM
      39S   11W   03   S
      42-12-17N   123-58-52W
      DUNITE   81 04

SWEDE CLAIM
      39S   11W   10   N
      42-11-44N   123-58-41W
      SERPENTINE   81 04
TAYLOR RIDGE PROSPECT
      41S   10W   04   "SW,SW"
      42-01-31N   123-53-42W   ULTRAMAFICS
TINCUP GROUP
      37S   10W   32   NW
      42-18-40N   123-54-45W
      "PYROXENITE, GABBRO"   81 04
TRAILS END
      31S   14W   26
      42-50-51N   124-19-09W
      SERPENTINE   81 04
TRIPLE ASBESTOS PROSPECT
      34S   12W   34
      42-35-29N   124-06-39W
UNCLE SAM (SLIDE CREEK)   38S   10W   11
      42-16-40N   123-50-20W
      CR   SERPENTINE
UPPER CHETCO AND SQUARE LAKE
      39S   10W   17
      42-09-53N   123-53-33W
      HARZBURGITE 81 04
UPPER CHETCO AREA PLACERS

38S   10W   18
42-15-47N   123-55-39W   AU   81 04

UPPER LAWSON AND HUNTLEY SPRING
37S   12W   07
42-22-51N   124-09-32W
LATERITES   81 04

UPPER PINES
38S   12W   11
42-16-52N   124-05-30W   CR

UPPER QUOSATANA PIT
36S   13W   33
42-24-52N   124-14-50W   STONE

US PLYWOOD
36S   14W   16
42-27-40N   124-22-01W
SAND & GRAVEL

WAHL SITE
32S   15W   17   CENTER
42-48-13N   124-30-04W
STONE

WAR BABY
32S   13W   23
42-46-10N   124-12-49W
METAVOLCANICS AND METASEDIMENTS

WEBFOOT
38S   10W   24
42-14-35N   123-49-00W

WEDDERBURN RANCH QUARRY
36S   14W   30   NE
42-26-04N   124-24-33W   STONE

WESTERN BUILDERS SUPPLY
32S   15W   11
42-48-50N   124-26-19W
SAND & GRAVEL

WILD HORSE MTN
36S   12W   18
42-27-42N   124-10-21W

WILDCAT GROUP
34S   12W   04   "SW,SW"
42-39-20N   124-08-35W

WILSON QUARRY
32S   15W   12   SE
42-48-54N   124-24-54W
SAND & GRAVEL

WINCHUCK PIT
     40S   12W   05
     42-01-43N   124-08-45W
     SAND & GRAVEL
WINCHUCK PIT
     41S   12W   05
     42-02-00N   124-08-14W   STONE
WINDY VALLEY GROUP
     37S   12W   32
     42-20-04N   124-08-56W
     SERPENTINE   78 11
WINTON MT. CHROME
     41S   11W   09
     42-00-53N   123-59-44W
     SERPENTINE   81 04
WONDER GROUP
     38S   10W   11   "SE,SE,SW"
     42-16-23N   123-50-43W
     CR   SERPENTINE

YELLOW JACKET
     38S   10W   23
     42-14-35N   123-50-39W

YELLOW MOON
     32S   10W   32   NW
     42-44-52N   123-54-06W
     METAVOLCANICS   81 04
YOUNG CLAIM
     38S   10W   02
     42-17-47N   123-50-30W   CR
YOUNG MINE
     38S   10W   35
     42-13-34N   123-50-43W
     AU   COUNTRY ROCKS ARE LAYERED SILICIFIED
     SANDSTONE AN

Josephine County

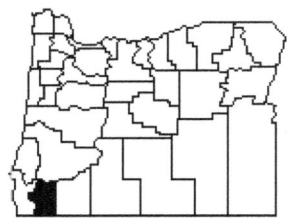

NAME
TOWNSHIP   RANGE   SECTION   SECTION
FRACTION
NORTH LATITUDE   WEST LONGITUDE
MINED MATERIAL
Last Update of Claim or Quarry if I was able to get it

A BAR/GOLD PAN PLACER MINE
    39S   07W   36   "SE,SE"
    42-07-47N   123-27-57W
ABBUY
    40S   09W   05
    42-07-26N   123-46-58W
ABEGG STOCKPILE SITE
    35S   06W   19
    42-31-02N   123-27-04W   STONE
ADITS AND SHAFT
    38S   05W   20   "W,SW"
    42-14-56N   123-19-34W
AJAX
    33S   08W   36   N
    42-39-38N   123-35-27W   AU   GREENSTONE
    81 02
AL BLASSON
    38S   09W   35   NW
    42-13-33N   123-44-12W
    SERPENTINE AND GREENSTONE
ALBERG
    40S   09W   05   NW
    42-07-21N   123-47-23W
    SERPENTINE   81 02
ALI
    40S   08W   24   "SE,SE"
    42-04-07N   123-34-60W   UHRAMAFICS
ALMEDA
    34S   08W   13   SE
    42-36-37N   123-35-03W
    AU CU AG PB DACITE PORPHYRY SLATE
ALTA
    36S   09W   23
    42-25-41N   123-43-49W
    SERPENTINE   81 01
ALTA
    39S   09W   02   SE
    42-12-10N   123-43-23W
    AU   SERPENTINE   81 02

ALTHOUSE
    40S   07W   22   NE
    42-04-46N   123-30-28W   CR
ALTHOUSE MINE
    41S   06W   12
    42-00-53N   123-21-57W
    AU   81 04

ALTHOUSE -RUN GULCH PLACER
     40S   07W   26
     42-03-25N   123-29-44W
     AU   GRAVELS   81 04
ANACONDA
     33S   05W   29
     42-40-48N   123-18-52W   AU
ANDERSON
     38S   08W   18   "NW,NE"
     42-16-17N   123-41-13W
     AU PT IR   CHANNEL AND BENCH GRAVELS
          81 04
ANDES & HOWARD PROSPECTS
     37S   05W   15   E
     42-20-55N   123-16-06W
     AU   "ARGILLITE, SANDSTONES, LIMESTONE"
          77 05
ANGORA QUARRY
     34S   07W   15
     42-36-55N   123-30-43W
     STONE
APEX OCCURRENCE.
     36S   08W   09
     42-27-15N   123-39-00W
     SERPENTINE   81 02
APPLEGATE AGGREGATES BAR   37S   02W   25
     42-19-00N   123-21-30W
     SAND & GRAVEL
APPLEGATE BAR PIT & PLANT
     36S   06W   30
     42-24-30N   123-27-02W
     SAND & GRAVEL

APPLEGATE RIVER
     37S   05W   20
     42-20-35N   123-19-04W
     SAND & GRAVEL
APPLEGATE-S OR CONCRETE
     36S   06W   20
     42-25-24N   123-26-24W
     SAND & GRAVEL
ARGO
     34S   08W   13
     42-37-18N   123-35-58W
     AU AG GREENSTONE 81 02

ARNOLD MINE
      41S   05W   16   "N,NE"
      42-00-20N   123-17-26W
      AU AG METASEDIMENTS   81 02
ARNOT
      36S   08W   12
      42-27-13N   123-35-25W

ARROWHEAD
      38S   06W   24   "NE,SE"
      42-14-56N   123-20-55W
      GREENSTONE 81 02
ASSOCIATED CHROMITE CLAIM
      39S   08W   13
      42-10-52N   123-35-17W
      CR   "SAXONITE, DUNITE" 81 02
ATLAS GOLD DREDGING
      40S   07W   04
      42-06-29N   123-31-23W
      AU PT   STREAM GRAVELS
BABCOCK
      38S   06W   27   "CENTER,W"
      42-14-13N   123-23-60W
      SERPENTINE   81 02

BABCOCK COPPER PROSPECT
      39S   06W   05
      42-12-11N   123-25-55W
      METABASALT GREENSTONE   81 02
BABYFOOT
      39S   09W   29
      42-08-46N   123-47-14W
BAILEY
      40S   08W   16
      42-05-00N   123-38-00W
      AU
BAILEY MINE/WALDO AREA PLACERS
      40S   08W   28   "S,SW"
      42-03-16N   123-39-14W
      AU PT   MODERN AND PLEISTOCENE GRAVELS
BAKER MINE
      33S   07W   07   SW
      42-42-50N   123-34-33W   GREENSTONE
      81 02
BEAR CAT CHROME
      41S   05W   16   SE
      42-00-08N   123-17-30W
      SERPENTINE   81 02

BEAR MINE
>38S   05W   28   "NW,NW"
>42-14-37N   123-18-26W
>METAVOLCANICS   81 02

BEAR PLACER
>38S   09W   36
>42-13-27N   123-42-20W
>AU PT NI FE   BENCH GRAVEL AND
>CONGLOMERATE   81 05

BECCA & MORNING GROUP
>38S   09W   07   "NW,SW"
>42-17-19N   123-49-28W
>AU   GREENSTONE-SERPENTINE   81 02

BENSON PLACER
>34S   07W   02   SE
>42-38-24N   123-29-12W
>AU   HIGH CHANNEL AND BENCH GRAVELS
>81 05

BENTON
>33S   08W   22   "SW,SE"
>42-40-35N   123-37-42W
>AU AG QUARTZ DIORITE   81 02

BETH PROSPECT
>34S   08W   28   "SW,NE"
>42-35-18N   123-38-48W
>AMPHIBOLITE 81 02

BIG BEAR
>36S   08W   35
>42-24-10N   123-36-36W
>CR   SERPENTINE   81 02

BIG BILL
>39S   09W   11   SW
>42-11-14N   123-43-58W
>SERPENTINE   81 02

BIG BUCK CLAIM
>37S   09W   28   "N,NW"
>42-19-48N   123-46-21W
>DUNITE   81 02

BIG CHIEF
>37S   08W   16
>42-21-23N   123-39-20W
>CR   SERPENTINE   81 02

BIG FOUR
>42-16-49N   123-47-59W
>38S   09W   07
>CR   SERPENTINE   81 02

BIG FOUR PLACER
>35S   07W   26   W
>42-29-51N   123-29-30W
>AU   HIGH BENCH GRAVEL 81 05

BIG SLIDE LODE
>33S   05W   10   "SE,SE"
>42-42-40N   123-16-09W
>GREENSTONE 81 02

BILLY BLUE
>37S   05W   01   CENTER
>42-22-58N   123-14-17W
>METASEDIMENTARY ROCKS   81 04

BLACK BEAR
>34S   08W   26   SW
>42-34-58N   123-36-59W
>AU AG   GREENSTONE-AMPHIBOLITE 81 02

BLACK BEAR CLAIM
>38S   09W   03   "SW,NE"
>42-17-45N   123-44-30W
>METAVOLCANICS   81 02

BLACK BEAR NO. 1 & 2
>41S   09W   07
>42-01-01N   123-48-37W
>"PERIDOTITE, DUNITE"   81 02

BLACK BEAUTY
>37S   09W   21   SE
>42-20-32N   123-46-21W

BLACK CHROME
>33S   05W   28
>42-40-27N   123-18-23W
>SERPENTINE 81 02

BLACK DIAMOND
>37S   09W   30
>42-19-34N   123-48-30W
>CR   DUNITE   81 02

BLACK DIAMOND
>40S   06W   31
>42-02-22N   123-27-44W
>CR   SERPENTINE 81 02

BLACK HAWK
>34S   08W   28   "E,NW"
>42-35-22N   123-39-11W
>AMPHIBOLITE 81 02

BLACK JACK
>35S   08W   03
>42-33-54N   123-38-25W
>AU   AMPHIBOLITE GNEISS 77 05

BLACK KING
      38S   09W   24   "CENTER,E"
      42-14-52N   123-42-20W
      CR   SERPENTINE   81 02
BLACK OTTER CLAIM
      37S   09W   20
      42-19-56N   123-46-42W
      CR   SERPENTINE   81 02
BLACK RABBIT & RIDGE CHROME
      33S   05W   11
      42-43-20N   123-15-28W
      SERPENTINE   81 02
BLACK ROCK CHROMITE
      37S   09W   09
      42-21-56N   123-46-08W
BLACK ROCK CHROMITE
      37S   08W   09
      42-21-55N   123-38-43W   CR   81 02
BLACK ROSE
      37S   09W   33
      42-18-27N   123-45-46W
BLACK STREAK
      39S   09W   24
      42-10-01N   123-43-02W
      CR   "SAXONITE, DUNITE" 81 02
BLACKHAWK
      35S   08W   04
      42-33-27N   123-38-34W
BLUE BELL
      34S   08W   30
      42-35-26N   123-41-43W
      SILICEOUS GREENSTONE   81 02

BLUE JAY CLAIM
      38S   05W   31   W
      42-13-17N   123-20-41W
      GREENSTONE (ANDESITE)   81 02
BLUE LODE
      38S   09W   35   NW
      42-13-36N   123-44-02W
      METAVOLCANICS   81 02
BLUE MULE CLAIM
      38S   05W   20   "SW,NW"
      42-15-11N   123-19-34W
      ANDESITE PORPHYRY 81 02
BLUE PRINCE CLAIM
      36S   09W   14   SE
      42-25-58N   123-43-20W
      CR   SERPENTINE   81 02

BLUEBUCKET #5
        41S   09W   08
        42-01-12N   123-47-21W
BLUFF PROSPECT
        37S   08W   32   NE
        42-18-46N   123-39-45W
        SERPENTINITE 81 02
BOLAN LAKE PLACER
        41S   07W   01   "E,E"
        42-01-48N   123-27-54W
BONANZA
        40S   06W   31   "NE,SW"
        42-02-48N   123-27-04W
        ULTRAMAFICS

BONE OF CONTENTION MINE
        38S   05W   25
        42-14-43N   123-14-24W
        AU   ARGILLITE   81 02
BOSWELL PROSPECT
        39S   07W   36
        42-07-50N   123-28-49W
        AU AG GREENSTONE 81 02
BOULDER CREEK MANGANESE & MINERAL LEDGE
        33S   05W   25
        42-40-45N   123-14-30W
BOWDEN PROSPECT
        39S   09W   08   N
        42-11-37N   123-46-30W
        GREENSTONE 81 02
BOWERS CHROME
        36S   09W   14
        42-26-14N   123-43-20W   CR
BRADBURY
        34S   08W   12   SW
        42-37-24N   123-35-43W
        AU   GREENSTONE 81 02
BRASS LEDGE QUARTZ CLAIM
        34S   08W   28
        42-35-27N   123-38-30W
        CU AU AMPHIBOLITE 81 01
BRASS NAIL PLACER
        34S   05W   24
        42-36-00N   123-14-20W

BRIGGS CREEK
        37S   09W   03
        42-22-52N   123-44-43W
        CR   SERPENTINE   81 02

BRIGGS CREEK PLACERS
         36S   08W   07
         42-27-15N   123-41-10W
         AU   CHANNEL AND TERRACE GRAVELS
BRIGGS POCKET
         41S   06W   14
         42-00-25N   123-22-45W   AU
         METASEDIMENTS   81 02
BRINER
         37S   05W   24
         42-20-25N   123-14-16W
BROOKLYN (GOLD PICK)
         40S   06W   30
         42-03-38N   123-27-16W
         ARGILLITE   81 02
BROWN
         40S   07W   36   "SE,NE"
         42-02-50N   123-28-06W
         METAVOLCANIC AND QUARTZITE   81 04
BROWN OWL MINE
         38S   05W   20   "SW,SE"
         42-14-38N   123-18-52W
         METAVOLCANICS   81 02
BROWN SCRATCH
         37S   09W   03   "NW,NE"
         42-23-13N   123-44-40W
         CR   "SAXONITE, DUNITE" 81 02

BROWN TOWN
         40S   05W   35
         42-03-05N   123-15-55W
BROWNTOWN SITE
         40S   07W   15   SW
         42-04-59N   123-30-51W   AU
BUCKEYE MINE
         36S   08W   25   "SE,NE"
         42-24-40N   123-34-55W
         GREENSTONE 81 02
BUCKHORN MINING CO.
         41S   09W   16
         42-01-00N   123-46-00W
BUCKSKIN
         38S   08W   19
         42-14-37N   123-41-00W   CR
BUFFALO GROUP
         34S   08W   21   "NW,SE"
         42-35-54N   123-38-46W
         GREENSTONE QUARTZITE SERPENTINE

BUHLER
       40S   07W   10
       42-06-32N   123-30-28W
BUNKER HILL
       35S   09W   02   SE
       42-33-16N   123-43-21W
       AU AG GREENSTONE 81 02
BURNS PROSPECT #2
       35S   08W   36
       42-28-58N   123-35-29W

BURROW PROSPECT
       35S   07W   20
       42-31-14N   123-33-05W
       SERPENTINE 81 02
BUSTER
       36S   09W   11
       42-26-52N   123-43-38W
       DUNITE   81 02
BUTTE
       37S   09W   30
       42-19-25N   123-48-24W
       PERIODOTITE 81 02
C C CHROME
       40S   07W   03   CENTER
       42-07-08N   123-30-46W
       SERPENTINE 81 02
CALIFORNIA GULCH
       33S   08W   26   SW
       42-40-19N   123-36-54W
       AU AG CU   METAGABBRO 81 02
CALIFORNIA MINE   33S
       07W   30   "NW,SW"
       42-40-23N   123-34-48W
       AU   GREENSTONE PORPHYRITIC ANDESITE
              81 01
CALIFORNIA OREGON
       35S   08W   02   "E,NW"
       42-33-39N   123-37-19W
       AU   TERRACE GRAVELS   81 05
CAL-ORE POWER CO.
       36S   05W   28
       42-24-37N   123-17-51W

CAMERON MINE
       40S   08W   34   SW
       42-02-45N   123-38-15W

CAMP BIRD CLAIM
     41S   06W   07   "NW,NW"
     42-01-24N   123-27-43W
     GREENSTONE 81 02

CANNON & CONNACHER PROSPECT
     39S   07W   34   "S,S,CENTER"
     42-07-44N   123-31-02W
     METASEDIMENTS   81 02

CANYON CK. CONS. GOLD MINES
     39S   09W   05   W
     42-12-19N   123-47-28W
     GREENSTONE 81 05

CARLTON GROUP
     35S   08W   10
     42-32-11N   123-38-03W
     GREENSTONE 81 02

CARNEGIE PLACE
     35S   08W   03   "NW,SE"
     42-33-22N   123-37-49W

CARTER CREEK PROSPECT
     39S   09W   15
     42-20-42N   123-45-13W
     SERPENTINE 81 02

CAT BIRD CLAIM
     40S   09W   28   E
     42-03-36N   123-45-41W

CATTY BUCK CLAIM
     36S   09W   22
     42-25-18N   123-44-26W
     CR   SERPENTINE 81 02

CAVYELL HORSE CREEK
     36S   09W   34
     42-23-36N   123-44-15W
     CR   SERPENTINE 81 02

CAVYELL HORSE MT.
     37S   09W   03
     42-23-19N   123-44-27W
     CR   DUNITE   81 02

CEDAR CREEK
     38S   08W   05   NW
     42-17-45N   123-40-25W
     SERPENTINE 81 02

CEDAR SPRINGS NICKEL DEPOSIT
     40S   10W   35
     42-02-43N   123-50-45W

CENTENNIAL
     35S   05W   26
     42-29-52N   123-15-44W
     GREENSTONE AND GRANITE 81 01

CHANCELLOR QUARRY
      35S   06W   13   SW
      42-31-25N   123-21-44W
      STONE

CHAPIN
      34S   05W   23   SE
      42-35-45N   123-15-10W
      ULTRAMAFICS 81 05

CHAPMAN CREEK
      39S   08W   14
      42-10-20N   123-36-27W
      CR   DUNITE   81 02

CHARLIE CANYON
      40S   09W   30
      42-03-35N   123-48-24W
      SERPENTINE   81 02

CHARLOTTE
      40S   05W   05
      42-07-34N   123-19-24W

CHATTY MINE
      38S   09W   26
      42-14-26N   123-43-40W
      AU   GREENSTONE 81 02

CHENEY CREEK QUARRY
      37S   06W   18   NW
      42-21-21N   123-27-31W
      STONE

CHETCO COPPER CO.GOLD PROSPECT
      39S   09W   10   SW
      42-11-09N   123-45-16W
      GABBRO AND SERPENTINE   81 04

CHETCO GROUP
      38S   09W   11
      42-16-40N   123-43-36W

CHOLLARD MINE
      40S   07W   17
      42-05-11N   123-33-09W
      CR   "SAXONITE, DUNITE" 81 02

CHROME CREST CLAIM
      35S   09W   36   NW
      42-29-23N   123-42-56W
      SERPENTINE   81 02

CHROME DOME
      41S   09W   09
      42-01-06N   123-46-23W
      DUNITE   81 02

CHROME FLAT
    36S   09W   14
    42-26-09N   123-43-42W
    CR   DUNITE   81 02
CHROME KING
    41S   09W   05   NE
    42-02-06N   123-46-45W
    SERPENTINE   81 02
CHROME KING NO. 1
    37S   10W   36
    42-18-39N   123-48-56W
    CR   SERPENTINE   81 02
CHROME KING NO. 3
    37S   10W   36
    42-18-45N   123-49-02W
    CR   SERPENTINE   81 02
CHROME OCCURRENCE
    34S   07W   22
    42-36-09N   123-30-43W
CHROME OCCURRENCE
    36S   08W   33
    42-23-45N   123-39-00W
    ULTRAMAFICS 81 03

CHROME OCCURRENCE
    38S   07W   26   NW
    42-14-13N   123-29-49W
CHROME OCCURRENCE
    38S   06W   18
    42-15-59N   123-27-17W
    ULTRAMAFICS 81 03
CHROME OCCURRENCE
    41S   07W   07
    42-01-03N   123-34-14W   81 03
CHROME PROSPECT
    34S   06W   33
    42-34-21N   123-24-55W
CHROME PROSPECT
    35S   07W   17
    42-31-17N   123-33-08W
CHROME PROSPECT
    35S   07W   34
    42-29-07N   123-30-46W
CHROME PROSPECT
    36S   09W   23   NE
    42-25-35N   123-43-10W
    SERPENTINE   81 03

CHROME PROSPECT
     36S   09W   23
     42-25-16N   123-44-04W
     ULTRAMAFICS 81 03

CHROME PROSPECT
     36S   08W   33
     42-23-45N   123-38-55W

CHROME PROSPECT
     37S   10W   36   N
     42-20-26N   123-49-32W
     ULTRAMAFICS   81 04
CHROME PROSPECT
     37S   09W   15
     42-21-18N   123-44-41W
     ULTRAMAFICS   81 03
CHROME PROSPECT
     37S   09W   22
     42-20-36N   123-44-23W
     ULTRAMAFICS   81 03
CHROME PROSPECT
     37S   09W   32   SE
     42-18-19N   123-46-59W
     SERPENTINE   81 03
CHROME PROSPECT
     37S   09W   32
     42-18-10N   123-45-56W
CHROME PROSPECT
     37S   09W   33   NW
     42-18-44N   123-46-09W
     SERPENTINE   81 03
CHROME PROSPECT
     38S   10W   02
     42-17-55N   123-50-13W   81 03
CHROME PROSPECT
     38S   10W   12
     42-16-24N   123-49-24W
CHROME PROSPECT
     38S   07W   22
     42-15-16N   123-30-30W
CHROME PROSPECT
     38S   07W   26   "E,W"
     42-14-11N   123-29-42W
     SERPENTINE   81 02
CHROME PROSPECT
     38S   06W   25
     42-14-09N   123-21-20W   CR

CHROME PROSPECT
    38S   06W   33
    42-13-17N   123-24-52W   CR
CHROME PROSPECT
    38S   05W   04
    42-17-18N   123-18-59W   CR
CHROME PROSPECT
    38S   05W   31   "SE,SE"
    42-13-04N   123-19-49W
    ULTRAMAFICS 81 03
CHROME PROSPECT
    38S   05W   36
    42-13-23N   123-14-19W   AU
CHROME PROSPECT
    39S   09W   06   "SW,NW"
    42-12-27N   123-41-38W
    SERPENTINE   81 03
CHROME PROSPECT
    40S   09W   21
    42-04-05N   123-45-26W
    ULTRAMAFICS 81 03

CHROME PROSPECT
    40S   07W   03   SE
    42-06-52N   123-30-21W
    METAVOLCANICS   81 04
CHROME PROSPECT
    41S   06W   16
    42-00-18N   123-24-45W   CR
CHROMIC OXIDE
    33S   05W   22
    42-40-54N   123-17-01W
CHROMITE OCCURRENCE
    34S   05W   19   W
    42-36-00N   123-20-29W   ULTRAMAFICS
    81 03
CHROMITE OCCURRENCE
    34S   05W   24
    42-36-01N   123-14-16W   ULTRAMAFICS
    81 03
CHROMITE OCCURRENCE
    34S   05W   30   NW
    42-35-21N   123-20-24W   SERPENTINITE
    81 03

206

CHROMITE RANCHERIE CREEK
    38S   09W   05   NW
    42-17-50N   123-47-23W
    SERPENTINE   81 04

CLEARWATER CHROME
    37S   09W   20   NW
    42-20-27N   123-47-34W
    SERPENTINE   81 02

CLEOPATRA CLAIM
    38S   09W   05
    42-17-41N   123-47-10W
    CR   SERPENTINE

COBALT GROUP
    36S   11W   33
    42-25-09N   124-00-23W

COHN LEDGE
    39S   07W   35
    42-08-01N   123-29-35W   AU   GREENSTONE
METASEDIMENTS   81 02

COLD SPRING
    35S   08W   09   NW
    42-32-58N   123-39-05W   AMPHIBOLITE
    81 02

COLUMBIA
    34S   05W   05
    42-38-40N   123-18-54W
    AU   CHANNEL GRAVELS   81 05

CONDORE
    39S   07W   34   SW
    42-07-46N   123-23-57W

CONTACT GROUP
    35S   07W   28
    42-30-06N   123-31-42W
    DARK METASEDIMENTARY ROCK   81 01

COPPER BELL
    35S   07W   28   NE
    42-30-06N   123-31-28W
    CU   GREENSTONE 81 01

COPPER BOY
    38S   09W   08
    42-16-50N   123-47-00W
    GREENSTONE 81 02

COPPER PROSPECT
    39S   06W   04   N
    42-12-03N   123-26-51W
    METAVOLCANICS   81 03

COPPER PROSPECT
    39S   06W   08   NW
    42-11-49N   123-26-31W
    METAVOLCANICS   81 03
COPPER QUEEN
    34S   06W   15
    42-37-11N   123-23-36W
    AU CU AG   GALICE FM   81 01
COPPER STAIN
    33S   08W   35   N
    42-39-48N   123-36-41W
    AU   GREENSTONE 81 02
COUGAR MINE
    33S   05W   22   NE
    42-41-30N   123-16-22W
    AU PORPHYRITIC GREENSTONE IN CONTACT WITH
    SERPENTINE   81 02
COURIER
    36S   09W   35
    42-14-12N   123-43-17W   AU
COWBOY MINE
    41S   08W   11   NE
    42-01-10N   123-36-29W
    CU AU AG ZN   SERPENTINE GREENSTONE   81 02

COX THOMPSON MINE
    37S   10W   36   S
    42-18-16N   123-49-34W
    SERPENTINE   81 02
COYOTE CREEK BAR
    33S   05W   29
    42-40-29N   123-19-02W
    SAND & GRAVEL

COYOTE MINE
    33S   05W   21   SE
    42-41-07N   123-17-41W   AU   GREENSTONE
    81 02
CRAMER
    35S   05W   18
    42-31-45N   123-20-38W   GREENSTONE
    81 02
CROWN
    37S   09W   28
    42-19-44N   123-45-40W
    CR   DUNITE (SERPENTINIZED)   81 02
CUTLER PLACE
    39S   09W   24
    42-09-57N   123-42-08W

CYNTHIA
    41S   05W   15
    42-00-34N   123-16-29W
    CR   SERPENTINE   81 02

DAISEY AND HAMMERSLEY
    34S   05W   14
    42-36-46N   123-15-10W
    AU AG GREENSTONE SERPENTINE
DAISY
    40S   08W   26   "SW,SW,SW"
    42-03-05N   123-37-35W

DARK CANYON
    38S   05W   19   CENTER
    42-14-54N   123-19-34W
    AU AG ARGILLITE   81 02
DARK STAR
    37S   10W   25
    42-19-13N   123-48-59W
    CR   PERIDOTITE   81 02
DEAD HORSE
    39S   06W   34   SW
    42-07-54N   123-23-57W
    SERPENTINE   81 01

DEAN & DEAN PLACER
    34S   08W   25   "N,SE"
    42-35-11N   123-35-12W
DEEP GORGE CHROMITE
    37S   09W   32
    42-18-45N   123-46-51W
    CR   OLVINE RICH PERIDOTITE (SERPENTINIZED)
        81 02
DEEP GRAVEL
    40S   08W   21   SE
    42-04-19N   123-38-54W
    AU   BUTCHER GULCH GRAVEL   81 05
DEER CREEK
    38S   07W   24
    42-14-45N   123-28-48W
DEER CREEK PROSPECT
    38S   07W   24   "SW,SW"
    42-14-44N   123-28-40W
    SERPENTINE   81 02

DEPRESSION BREAKER
    34S   05W   29
    42-35-03N   123-19-07W   AU

DEWEY
     33S   05W   22
     42-41-40N   123-16-07W
DICK
     35S   05W   08   "W,NE"
     42-32-52N   123-18-51W
     AU   METAGABBRO GREENSTONE QUARTZITE
          81 02
DICK AND DICK REYNOLDS
     40S   09W   28
     42-02-26N   123-45-30W
     CR   PERIDOTITE   81 02
DIRTY FACE
     37S   09W   29
     42-19-31N   123-47-27W
     CR   PERIDOTITE   81 02

DO LITTLE CHROME
     41S   09W   04
     42-01-26N   123-46-06W
DOROTHEA
     33S   05W   22   NW
     42-41-04N   123-17-03W
     AU AG GABBRO   81 02
DOROTHY CHROME
     33S   05W   13
     42-42-01N   123-13-58W
     CR   SERPENTINE   80 12
DOTTIE MAY
     39S   08W   30
     42-09-10N   123-41-48W
     DUNITE   81 02
DRY DIGGINS
     36S   05W   14   SW
     42-26-21N   123-15-43W
     AU AG   81 02
DUNBAR PLACER
     36S   11W   13
     42-27-26N   123-57-00W   AU
DUNN CREEK
     41S   06W   18
     42-00-23N   123-27-12W
DUTCH LEDGE
     38S   05W   10   "NW,SW"
     42-16-42N   123-16-57W
     AU   ARGILLITE AND CHLORITE ALBITE SCHISTS

EAGLE
      35S   05W   06   SW
      42-33-09N   123-20-37W
      ARGILLITE   81 02
EARLY SUNRISE
      37S   10W   35
      42-18-09N   123-50-22W
      CR   SERPENTINE   81 02
EASTMAN NO. 1
      33S   05W   35
      42-39-53N   123-15-58W
EGGERS AND HANCE
      41S   09W   15
      42-00-10N   123-44-55W
EIGHT DOLLAR MOUNTAIN
      38S   08W   21
      42-14-51N   123-39-04W
      NI   93 04
EIGHT DOLLAR NO. 1
      38S   08W   29   "S,NE"
      42-14-18N   123-39-49W
      CR   BLOCKY PERIODOTITE
ELDER CREEK
      40S   07W   30   SW
      42-03-24N   123-34-37W
      SERPENTINE   81 02
ELDER MANGANESE
      39S   05W   06   S
      42-12-22N   123-20-21W
      GEMSTONES(RHODONITE)   METASEDIMENTS
      81 02
ELEPHANT
      40S   07W   18   SE
      42-05-06N   123-33-54W   GREENSTONE
      81 02
ELKHORN CREEK MANGANESE
      36S   09W   13
      42-26-46N   123-42-32W
      "AMPHIBOLITE GNEISS,
      BANDED QUARTZITE" 81 01
ELKHORN CREEK MINE
      36S   09W   13
      42-25-50N   123-42-58W
      CR   DUNITE   81 02

ELKHORN PLACER MINE
    36S   09W   24   S
    42-25-23N   123-42-25W
    AU   CHANNEL AND BENCH GRAVELS
EMPIRE
    36S   07W   03   NW
    42-28-21N   123-30-50W
    MASSIVE AND SCHISTOSE GREENSTONES
ENTERPRISE PLACER
    41S   07W   16
    42-01-00N   123-32-00W   AU
"ERRIE, CLARENCE, ETC."
    40S   09W   33
    42-02-21N   123-45-31W   HARZBYRGITE
    81 02
ESTERLEY CHROME MINE
    40S   08W   22
    42-04-46N   123-37-39W
    CR   SERPENTINE   81 02
ESTERLEY PLACER   40S   08W   10   42-05-
53N   123-38-12W   AU PT AG   LLANO DE ORO FM.
    81 04
EUREKA MINE
    37S   09W   22   "NE,NW"
    42-20-42N   123-44-56W
    AU AG GREENSTONE 81 02
EVA UNDERHILL
    35S   08W   08   "SW,NW"
    42-32-37N   123-40-40W
EXCHEQUER MINE
    37S   05W   34
    42-18-34N   123-15-54W
    AU AG GREENSTONE ARGILLITE   81 02
FALL CREEK COPPER
    38S   09W   04
    42-17-44N   123-46-03W
    CU AU AG   GREENSTONE AND SERPENTINE
        81 04
FANDORA MINE
    33S   05W   22   "W,SE"
    42-41-03N   123-16-35W   METABASALT
SERPENTINE   81 02

FEBRUARY CLAIM
    37S   09W   33
    42-18-21N   123-45-34W
    SERPENTINE   81 02

FLANAGAN
  35S   07W   35
  42-28-57N   123-30-05W
     AU   TERRACE GRAVEL   81 05
FLANAGAN MINE
  36S   07W   02
  42-28-21N   123-30-01W   AU
FLORENE MINE
  33S   05W   21   SE
  42-41-07N   123-17-21W
     METAVOLCANICS SERPENTINE   81 02
FOSTER ASBESTOS
  38S   09W   36
  42-13-04N   123-43-04W
     SERPENTINE   81 02

FOUR CORNERS SITE
  40S   08W   26   NE
  42-03-49N   123-37-08W   STONE
FOUR LEAF CLOVER PLACER
  40S   06W   19
  42-04-35N   123-27-32W
FOUR STAR PLACER
  40S   07W   13   "SE,NE"
  42-05-27N   123-27-56W
FOWLER GROUP
  35S   08W   10   "NE,SW"
  42-32-27N   123-38-02W
FRECKLES
  36S   07W   32
  42-23-26N   123-33-08W
FREE AND EASY
  38S   08W   32   S
  42-12-60N   123-42-39W
     LATERITES
FREE AND EASY LATERITE
  38S   08W   32   S
  42-13-05N   123-40-16W
     HARZBERGITE PLUS OR MINUS SERPENTINE 81 02
FREEMAN PIT
  40S   08W   17
  42-05-29N   123-40-40W
     SAND & GRAVEL
FROG POND
  41S   07W   15
  42-00-21N   123-30-43W
     AU AG METASEDIMENTARY ROCKS DIORITE
     SERPENTINE   81 02

FRY GULCH MINE
        40S   08W   28   N
        42-03-40N   123-39-01W
        AU   81 05
FRYE BAR
        36S   06W   20
        42-25-37N   123-26-18W
        SAND & GRAVEL
GALICE CHIEF
        35S   08W   16
        42-31-56N   123-38-53W
GALLAGHER CHROME
        37S   09W   03
        42-22-34N   123-44-54W
        CR   DUNITE   81 02
GEM QUARTZ MINE
        39S   07W   36   "NE,SE"
        42-07-59N   123-27-59W
        AU   ANDESITE   81 02
GEO MCALLISTER
        36S   07W   05   S
        42-28-00N   123-33-11W
        METAVOLCANICS   81 04
GLENDA LOU
        37S   05W   11
        42-22-26N   123-15-29W   GREENSTONE
        81 02
GODS LITTLE AREA
        40S   09W   32   SW
        42-02-30N   123-46-56W

GOFF MINE
        33S   07W   29   "NW,NW"
        42-40-49N   123-33-37W
        METAVOLCANICS   81 02
GOFF PLACER
        33S   07W   33
        42-39-40N   123-31-30W   AU
GOFF PLACER
        34S   06W   05   "N,NW"
        42-38-57N   123-26-21W
        AU   TERRACE GRAVELS   81 05
GOLD BLANKET
        37S   09W   04
        42-22-39N   123-46-27W   AU

GOLD BLANKET MINE
>   38S   09W   14   SW
>   42-16-05N   123-44-01W   GREENSTONE
>   81 02

GOLD BOND
>   37S   09W   05
>   42-22-57N   123-46-37W   AU

GOLD BUG
>   33S   08W   26   NE
>   42-40-46N   123-36-19W
>   AU   GREENSTONE 81 02

GOLD BUTTE PROSPECT
>   37S   10W   36
>   42-18-47N   123-48-59W
>   DUNITE   81 02

GOLD CHIEF
>   36S   05W   03   CENTER
>   42-28-12N   123-16-45W
>   AU   METAVOLCANICS   81 02

GOLD COIN
>   33S   05W   22
>   42-40-59N   123-17-14W

GOLD CUP PLACER
>   34S   05W   02   "E,NW"
>   42-38-48N   123-15-27W

GOLD MINE #1
>   33S   07W   33
>   42-39-26N   123-32-03W   AU

GOLD MINE #2
>   33S   06W   24   NE
>   42-40-55N   123-20-52W   AU

GOLD MINE #3
>   33S   06W   24
>   42-40-59N   123-21-39W   AU

GOLD MINE #4
>   33S   05W   14
>   42-41-48N   123-16-23W   AU

GOLD MINE #5
>   34S   05W   07   "SE,SE"
>   42-37-32N   123-19-45W   AU

GOLD MINE #6
>   35S   08W   07   "SW,NW"
>   42-32-41N   123-41-52W
>   AU   METAGABBRO 81 03

GOLD MINE #7
>   35S   08W   09
>   42-32-26N   123-38-41W
>   AU   93 04

GOLD MINE #8
        36S   05W   26
        42-24-34N   123-16-36W   AU
GOLD MINE #9
        38S   09W   02   SW
        42-12-54N   123-44-05W   AU
GOLD MINE   #10
        39S   09W   11   NE
        42-11-40N   123-43-29W   AU
GOLD MINE #11
        40S   08W   09   NE
        42-06-26N   123-39-08W   AU
GOLD MINE #12
        40S   08W   22   SW
        42-04-25N   123-38-09W   AU
GOLD MINE #13
        40S   07W   01
        42-07-02N   123-28-31W   AU
GOLD MINE #14
        40S   07W   01
        42-07-04N   123-28-31W   AU
GOLD MINE #15
        40S   07W   01   SE
        42-07-04N   123-28-28W   AU
GOLD MINE #16
        40S   07W   01   NE
        42-07-26N   123-28-00W   AU
GOLD MINE #17
        40S   07W   03   SW
        42-06-41N   123-30-44W   AU
GOLD MINE #18
        40S   07W   03
        42-06-47N   123-31-21W   AU

GOLD MINE #19
        40S   07W   27
        42-03-32N   123-31-08W   AU
GOLD MINE #20
        40S   05W   33
        42-02-27N   123-17-59W   AU
GOLD MINE #21
        40S   08E   26   "SE,SW"
        42-03-36N   123-37-16W   AU
GOLD NOTE
        33S   05W   25
        42-40-21N   123-13-37W
GOLD OCCURRENCE #1
        35S   09W   27
        42-30-06N   123-44-52W
                    216

GOLD OCCURRENCE #2
	35S	08W	28	NW
	42-30-04N	123-39-06W
GOLD PLACER 1
	34S	08W	28	"N,NE"
	42-35-31N	123-38-43W
GOLD PLACER 2
	34S	07W	03	"NW,SE"
	42-38-43N	123-30-40W
GOLD PLACER PROSPECT 1
	33S	07W	25
	42-40-19N	123-27-58W
GOLD PLACER PROSPECT 2
	33S	07W	36
	42-39-37N	123-28-22W

GOLD PLACER PROSPECT 3
	34S	08W	22
	42-35-39N	123-38-12W
GOLD PLACER PROSPECT 4
	38S	09W	25
	42-13-45N	123-43-01W
GOLD PLACER PROSPECT 5
	38S	09W	33
	42-13-28N	123-45-37W
GOLD PLATE MINE
	35S	08W	04	NW
	42-33-30N	123-39-26W
	AU	ALTERED VOLCANICS TO AMPHIBOLE
	GNEISS OF ROGUE FM 81 02
GOLD PROSPECT 1
	33S	08W	17
	42-41-21N	123-40-42W
GOLD PROSPECT 2
	34S	08W	01	"SW,SE"
	42-38-18N	123-35-19W
	METAVOLCANICS	81 04

GOLD PROSPECT 3
	34S	08W	23	"SE,NW"
	42-36-14N	123-36-50W
	METAVOLCANICS	81 04
GOLD PROSPECT 4
	34S	08W	26
	42-35-13N	123-36-20W
GOLD PROSPECT 5
	34S	08W	36	NE
	42-34-32N	123-35-31W

GOLD PROSPECT 6
     35S   08W   17   "SW,SW"
     42-31-21N   123-40-44W   AMPHIBOLITE
     81 03
GOLD PROSPECT 7
     35S   07W   10   "NE,NE"
     42-32-55N   123-30-19W   GREENSTONE
     81 03
GOLD PROSPECT 8
     36S   09W   05   NW
     42-28-18N   123-47-24W

GOLD PROSPECT 9
     36S   09W   26   "S,NW"
     42-24-45N   123-44-00W
     DIORITE   81 03
GOLD PROSPECT 10
     36S   08W   06
     42-27-42N   123-41-40W
GOLD PROSPECT 11
     36S   07W   05   "S,S"
     42-27-45N   123-33-12W
     METAVOLCANICS   81 04
GOLD PROSPECT 12
     36S   07W   22   "SW,SW"
     42-25-07N   123-31-21W
     METAVOLCANICS   81 04
GOLD PROSPECT 13
     37S   06W   10   "NW,NW"
     42-22-25N   123-24-11W
     METAVOLCANICS   81 03

GOLD PROSPECT 14
     37S   05W   28   "SW,NW"
     42-19-37N   123-18-17W
     METAVOLCANICS   81 03
GOLD PROSPECT 15
     37S   05W   33   SW
     42-18-21N   123-18-09W
     METAVOLCANICS   81 03
GOLD PROSPECT 16
     38S   09W   10   "E,NE"
     42-16-57N   123-44-16W   GREENSTONE
     81 04
GOLD PROSPECT 17
     38S   09W   11   "SW,NW"
     42-16-50N   123-44-06W   GREENSTONE
     81 04

GOLD PROSPECT 18
    38S   09W   22   SE
    42-14-38N   123-44-30W   GREENSTONE
GOLD PROSPECT 19
    40S   07W   31
    42-03-04N   123-34-09W
    METASEDIMENTS   81 04
GOLD PROSPECT 20
    40S   06W   06   "W,NW"
    42-07-22N   123-27-50W
GOLD PROSPECT 21
    40S   06W   19   "NW,NW,NW"
    42-04-50N   123-27-40W
    METASEDIMENTS   81 03
GOLD PROSPECT 22
    41S   07W   03   "NW,SW"
    42-01-43N   123-31-18W
    METAVOLCANICS   81 04

GOLD PROSPECT 23
    41S   07W   10   "SW,SE"
    42-00-35N   123-30-47W
    METAVOLCANICS AND METASEDIMENTS
GOLD PROSPECT 24
    41S   07W   13   "N,NW"
    42-00-35N   123-28-48W
    METASEDIMENTS   81 04
GOLD PROSPECT 25
    41S   07W   13   W
    42-00-12N   123-28-41W
GOLD PROSPECT 26
    41S   07W   14   SE
    42-00-07N   123-29-05W   81 03
GOLD RIDGE
    38S   09W   14   SW
    42-16-15N   123-44-03W   AU
    METAVOLCANICS SLATE   81 02
GOLD STAR MINE
    35S   05W   27   "NE,NW"
    42-30-11N   123-16-46W
    METAVOLCANICS   81 02
GOLDEN GLOW
    33S   05W   15
    42-41-47N   123-17-02W   AU
GOLDEN LIGHT CLAIM
    33S   07W   31   SE
    42-39-23N   123-33-56W   AU

GOLDEN MARY PROSPECT
      36S   05W   34   "N,N"
      42-24-12N   123-16-35W
      AU AG METATUFF RHYOLITE 81 02

GOLDEN PHEASANT GROUP
      35S   08W   03
      42-33-39N   123-37-43W   GREENSTONE
      81 02
GOLDEN PRINCESS
      38S   08W   30   "SW,SW"
      42-14-26N   123-40-46W
      AU   BENCH GRAVELS   81 05
GOLDEN RING MINE
      33S   05W   14   "SE,NW"
      42-42-12N   123-15-43W
      METAVOLCANICS SERPENTINE   81 02
GOLDEN WEDGE
      34S   08W   14   "S,S"
      42-36-37N   123-36-43W   AU   GREENSTONE
      81 02
GOOD FRIDAY
      37S   09W   32
      42-18-49N   123-47-03W
      CR   DUNITE   81 02
GOPHER MINE
      35S   05W   08   "SE,SE"
      42-32-11N   123-18-31W
GRANITE HILL MINE
      35S   05W   26   SW
      42-29-41N   123-15-49W
      AU AG QUARTZ DIORITE   81 02
GRANITE HILL PLACER
      35S   05W   27
      42-29-04N   123-16-09W   AU

GRAVE CREEK PLACER
      34S   06W   09
      42-38-09N   123-24-24W
      AU AG CHANNEL GRAVELS   81 05
GRAVES CREEK CHROME
      34S   05W   06
      42-38-57N   123-20-23W
      CR   SERPENTINE 81 02
GRAY BUCK GROUP GROUP
      37S   09W   16
      42-21-02N   123-46-07W
      CR   81 02

GRAY ROCK
    38S   09W   35   S
    42-13-04N   123-43-37W
    SERPENTINE   81 02
GRAY ROCK EXTENSION
    37S   05W   13   E
    42-21-11N   123-13-55W
    METAVOLCANICS
GRAYBACK PROSPECT
    39S   06W   35
    42-08-07N   123-22-26W
    META-ANDESITE   81 02
GREENBACK
    33S   05W   32   S
    42-39-13N   123-18-23W
    AU AG GREENSTONE 81 02
GREENBACK CHROME
    33S   05W   34   "NE,NW"
    42-40-04N   123-16-50W
    SERPENTINE

GRIFFIN CHROMITE
    38S   08W   18
    42-16-06N   123-41-19W
    CR   SERPENTINE   81 02
GRIZZLY MINE
    40S   06W   33   "W,SW"
    42-02-34N   123-25-27W

GROUSE MOUNTAIN
    36S   05W   27
    42-25-00N   123-17-05W
GRUBSTAKE
    34S   08W   13
    42-37-00N   123-35-20W
GUNNEL ROAD
    36S   07W   14
    42-26-39N   123-29-35W   STONE
GYPSUM OCCURRENCE
    35S   08W   02
    42-33-12N   123-36-19W
HAMLIN
    36S   07W   07   NW
    42-27-25N   123-34-40W
    SERPENTINE   81 02
HAMMERSELY MINE
    34S   05W   30   NW
    42-34-57N   123-20-37W
                SERPENTINE   81 02
221

HANSEN
       37S   09W   29
       42-19-24N   123-47-22W
       CR   SERPENTINE   81 02
HANSEN MINE
       35S   09W   13   "NE,NE"
       42-32-08N   123-42-03W   AMPHIBOLITE
METAGABBRO 81 02
HAP CLAIMS
       37S   09W   10   "SE,NW"
       42-22-06N   123-44-50W
       CR   81 02
HAPPY CAMP
       40S   07W   30
       42-03-49N   123-34-37W
       CR   SERPENTINE   81 02
HAPPY DAY GROUP
       41S   07W   13
       42-00-35N   123-28-48W
       AU AG METAVOLCANICS   81 02
HARD-TO-GET CLAIM
       36S   09W   22
       42-25-13N   123-44-39W
       SERPENTINE   81 02
HAROLD CHROME
       33S   05W   29
       42-40-09N   123-19-03W
HAVENS
       36S   07W   10
       42-27-18N   123-31-02W

HAYDEN
       34S   06W   13   E
       42-36-55N   123-21-02W
       AU AG GREENSTONE 81 02
HELLGATE PLACER
       35S   07W   10   "W,SW"
       42-32-23N   123-30-19W
       AU   TERRACE GRAVELS   81 05

HENDERSON PIT
       35S   07W   06   SE
       42-33-18N   123-33-55W
       SAND & GRAVEL

HERCULES GROUP
    34S   07W   06
    42-38-57N   123-33-45W
    BLACK SHALE SLATE 81 02

HIBBARD PROSPECT
    35S   07W   23   SW
    42-30-32N   123-29-52W
    AU AG ROGUE VOLCANICS   81 02

HIDDEN TREASURE
    36S   05W   27   NE
    42-24-56N   123-16-46W   DECOMPOSED
TONALITE   81 02

HIGH BUSHY CLAIM
    41S   05W   16
    42-00-21N   123-17-19W
    SERPENTINE   81 02

\HIGH GAP CHROME
    41S   05W   16   NW
    42-00-29N   123-18-18W
    SERPENTINE   81 02

HIGH GRAVEL
    40S   08W   33   NE
    42-02-60N   123-38-32W
    AU PT   WEATHERED TERRACE GRAVELS AND
    COLLUVIUM   81 05

HIGH RIDGE CLAIM
    37S   09W   29
    42-19-36N   123-47-44W
    DUNITE DIKES 81 02

HIGH VIEW PROSPECT
    37S   10W   24
    42-20-16N   123-49-08W
    PERIDOTITE   81 02

HILL PROSPECT
    35S   06W   06
    42-34-05N   123-27-60W   SLATE

HINKLE LAKE CLAIM
    41S   05W   10
    42-01-22N   123-16-49W
    QUARTZITE   81 03

HOLTON CREEK LIMESTONE
    39S   08W   14   "NE,NE"
    42-10-57N   123-36-09W

HORSE SHOE
    40S   07W   17   "N,NE"
    42-05-42N   123-32-49W
    AU AG GREENSTONE 81 03

HORSE SHOE LODE
        37S   08W   32
        42-18-19N   123-38-55W
        SERPENTINE   81 03
HORSEHEAD PLACER
        38S   05W   21   "S,S"
        42-14-45N   123-17-24W
        AU   COLLUVIUM AND CREEK GRAVELS   81 05
HORSESHOE LODE
        33S   05W   28   NW
        42-40-44N   123-18-14W
        AU AG GREENSTONE 81 03
HOWLAND
        35S   09W   24   SW
        42-30-34N   123-42-53W
        AU AG AMPHIBOLITE METAGABBRO SERPENTINE
        81 03
HUGHES CHROME MINE #2
        38S   06W   34
        42-13-02N   123-24-16W
HUGO SILICA
        35S   06W   05   NE
        42-33-20N   123-25-59W
        SILICA(LODE) QUARTZ DIORITE   81 03
HUMDINGER
        38S   05W   21   N
        42-16-02N   123-17-55W
        AU AG GREENSTONE ARGILLITE   81 03
HYDE BAR
        37S   05W   35   "N,CENTER"
        42-18-57N   123-15-49W
        SAND & GRAVEL   93 04
HYDE BAR AND PLANT
        37S   05W   35
        42-18-17N   123-15-56W
        SAND & GRAVEL
HYTEMPERATURE GROUP
        37S   08W   24
        42-20-00N   123-36-00W   CLAY
IDA MINE
        35S   05W   26   SE
        42-29-45N   123-15-08W
        AU AG GREENSTONE 81 03
INDEPENDENCE
        34S   08W   13
        42-36-32N   123-35-25W
        AU   GREENSTONE 81 03

INDEPENDENCE PLACER
  38S   08W   30   NE
  42-14-01N   123-41-20W
  AU   LOWER-UPPER BENCH GRAVEL AND
  CONGLOMERATE   81 05
INMAN
  38S   09W   02
  42-17-49N   123-43-33W
IRON GROUP
  41S   07W   02
  42-01-31N   123-29-43W
  ARGILLITE
IRON HAT GROUP
  37S   05W   17   SE
  42-21-04N   123-18-32W
  METAVOLCANIC   81 03

IRON MINE
  40S   06W   29   NW
  42-09-11N   123-20-44W
  ARGILLITE
IVERSON & POHLMAM PLACER
  34S   06W   08
  42-37-35N   123-26-37W
J D PLACER
  35S   08W   02
  42-33-51N   123-36-12W   AU
J. & B. EXCAUTING
  36S   05W   26
  42-24-37N   123-15-45W   STONE
J.C.L. MINE
  33S   08W   35   W
  42-39-29N   123-37-10W
  AU AG GREENSTONE AMPHIBOLE SCHIST   81 03
JACK SHADE CHROMITE
  37S   09W   21
  42-19-56N   123-45-33W
  CR   DUNITE AND SERPENTINE   81 04
JACKIE
  33S   05W   22   SW
  42-41-01N   123-16-51W
  SERPENTINE   81 05

JACKPOT #2
  41S   07W   05   NW
  42-02-11N   123-33-25W
CHERT 81 03

JACKPOT MINE

225

```
               33S      05W      20       "SW,NW"
               42-41-17N         123-19-29W
               METASEDIMENTS          81 04
JACKS CREEK
               34S      05W      28
               42-34-56N         123-18-43W      AU
JANTZER
               33S      05W      23
               42-41-01N         123-15-45W      81 03
JEWETT MINE
               36S      05W      27       "SW,SW"
               42-24-15N         123-16-52W
               AU AG  METAVOLCANICS         81 03
JIM BUS
               37S      09W      21
               42-20-27N         123-46-23W
               CR      DUNITE          81 03
JIM FISHER CHROME
               36S      09W      34
               42-23-51N         123-45-09W
JOHN HALL GROUP
               34S      05W      18       NE
               42-37-03N         123-19-53W
               AU      GREENSTONE  81 03
JOHNSON PLACER
               41S      07W      09
               42-00-58N         123-31-53W
JOKER
               34S      08W      23
               42-36-00N         123-36-34W
JONES MARBLE DEPOSIT
               38S      05W      31       NE
               42-13-19N         123-20-18W         STONE(LIMESTONE)
               METASEDIMENTS          81 03

JOSEPHINE CREEK
               39S      09W      26
               42-04-15N         123-43-22W
               NI      SERPENTINE  81 03
JOSEPHINE CREEK PLACERS
               38S      09W      36
               42-13-04N         123-42-53W
               AU PT NI         TERRACE AND CHANNEL GRAVELS
                    81 05
JOSEPHINE NO. 4 CLAIM
               39S      09W      12
               42-10-60N         123-42-21W
               CR      SERPENTINE  81 02
JULIANO
```

```
                    35S     07W     32
                    42-28-50N        123-33-29W
JUMP OFF JOE CREEK SITE
                    35S     06W     20
                    42-31-08N        123-25-38W        SAND & GRAVEL


JUMPOFF JOE CREEK PLACERS
                    34S     05W     24
                    42-36-26N        123-14-16W
                    AU      CHANNEL AND BENCH GRAVELS
JUNE GROUP
                    40S     07W     23
                    42-04-30N        123-29-44W                CHERT
KELLEY'S PIT
                    36S     06W     08
                    42-26-59N        123-26-29W
                    SAND & GRAVEL
KERBY QUEEN MINE
                    40S     07W     17      N
                    42-05-21N        123-33-05W
                    CU      METAVOLCANIC        81 03
KERBY SITE
                    39S     08W     04      SW
                    42-11-57N        123-39-24W        SAND & GRAVEL
KESTER MINE
                    40S     06W     29      NE
                    42-03-53N        123-25-42W
KEYSTONE GROUP
                    33S     08W     34
                    42-39-15N        123-38-13W                GREENSTONE
SERPENTINE  SERPENTINE
KING MINE
                    38S     05W     17      SE
                    42-15-48N        123-18-46W
                    METAVOLCANICS METASEDIMENTS  81 03
KING MOUNTAIN PIT
                    33S     05W     24
                    42-41-29N        123-14-36W        STONE
KING TUTT
                    34S     05W     29      "SE,NW"
                    42-35-18N        123-19-16W
                    AU      DIORITE         81 03
KLONDIKE MINE
                    35S     05W     22      SW
                    42-30-38N        123-16-57W
                    AU      GREENSTONE  81 03

KLUM PLACER MINE
```

34S   07W   01
42-39-00N   123-28-22W
AU   BENCH AND CHANNEL GRAVELS

KNAPPE GULCH
37S   09W   33
42-18-38N   123-46-11W

KNAPPE POINT
37S   09W   33
42-18-45N   123-46-14W

L.E.J. ASBESTOS
37S   06W   09
42-22-05N   123-24-53W   ASBESTOS
SERPENTINIZED PERIDOTITE 81 03

LAMBTONGUE
35S   05W   17   "NE,NE"
42-32-03N   123-18-28W
AU AG METAGABBRO 81 03

LARKSPUR
41S   05W   02
42-01-39N   123-15-48W
QUARTZITE   81 03

LAST CHANCE
37S   09W   15
42-21-04N   123-44-24W
SERPENTINE AND DIORITE   81 02

LAST CHANCE
39S   08W   19
42-09-24N   123-41-26W
CR   SAXONITE (PERIDOTITE)   81 03

LAST CHANCE PLACER
34S   08W   26   "SE,SE"
42-34-53N   123-36-07W

LAST DRINK NO. 1
41S   09W   07   "E,E"
42-00-48N   123-47-41W
SAXONITE PYROXENITE   81 02

LAWRENCE MINE
33S   07W   19   "SW,SW"
42-40-57N   123-34-38W
AU   GREENSTONE SERPENTINE

LAZY DAZE MINE
33S   05W   10
42-42-52N   123-16-37W   AU

LAZY MAN
33S   05W   14
42-41-51N   123-15-58W

LEANING PINE PROSPECT
      39S   09W   22
      42-20-16N   123-45-17W
      SERPENTINE 81 03
LEGAL TENDER MINE
      34S   08W   11   "NW,SW"
      42-37-46N   123-37-19W
      METAVOLCANICS SERPENTINE QUARTZITE 81 03
LEIPOLD PLACER
      35S   08W   03   "SE,SE"
      42-33-06N   123-37-27W
      AU   CHANNEL GRAVELS   81 05
LELAND (SBK) PLACER
      34S   06W   09   N
      42-38-12N   123-24-53W
LEONARD PLACER
      40S   07W   04   "SE,SW,SW,SE"
      42-06-55N   123-31-49W
LEWIS PLACER
      34S   06W   06
      42-38-41N   123-27-11W
LIGHTNING RIDGE
      36S   09W   24   "N,N"
      42-25-45N   123-42-59W   "QUARTZITE,
      MICA SCHISTS, AMPHIBOLITES"
LILLY
      40S   08W   35   SE
      42-02-20N   123-36-15W   CU   GREENSTONE
      81 03
LIMESTONE OCCURRENCE
      38S   06W   06   "NE,SE"
      42-17-35N   123-26-49W
      METASEDIMENTS   81 03
LINDH & WEBBER
      35S   07W   23   SE
      42-30-36N   123-29-19W
      AU   STREAM GRAVELS   81 05
LITTLE GEM
      39S   07W   36   W
      42-07-56N   123-28-36W
      METAVOLCANICS   81 03
LITTLE MARVEL
      35S   05W   03   "NW,NE"
      42-33-37N   123-16-38W
      AU   GRAVEL   81 05

LITTLE PICKETT CREEK
         35S   07W   22
         42-30-46N   123-30-58W   GREENSTONE
         81

LITTLE SILVER
         36S   10W   01   NW
         42-28-22N   123-49-55W   METAGABBRO
         81 03
LLANO DE ORO PIT NO. 2
         40S   08W   22   NW
         42-04-43N   123-38-16W   AU   81 03
LOGAN'S SAILOR GULCH PLACER
         40S   08W   27
         42-03-16N   123-37-25W
         AU   TERRACE GRAVELS AND COLLUVIUM
                 81 05
LONE EAGLE
         36S   09W   01   "NW,NW"
         42-28-41N   123-43-06W   ULTRAMAFICS
         81 03
LONE QUEEN LEDGE
         37S   08W   02
         42-23-13N   123-36-49W
LONE STAR PLACER
         40S   07W   28   "S,S"
         42-03-11N   123-31-32W
         VERY IRREGULAR PORPHYRY AND SHALE BEDROCK
LONNON ROAD
         36S   06W   36   SW
         42-23-39N   123-21-48W
         STONE(GRANITE)

LOOKOUT OCCURRENCES
         39S   09W   01   "S,S"
         42-11-48N   123-42-43W
         SERPENTINE   81 03
LOONEY MINE
         33S   08W   14   SE
         42-41-56N   123-36-29W
         HORNBLENDE GABBRO LOCALLY RECRYSTALLIZED
LOST FLAT
         35S   08W   17   NE
         42-31-59N   123-39-44W   AMPHIBOLITE
         81 03
LOST PROSPECT
         41S   06W   02
         42-01-54N   123-22-33W
         AU   META-ANDESITE   81 03

LOUISE CREEK PIT
   35S   06W   20
   42-31-11N   123-25-43W
   SAND & GRAVEL
LOW BOY CLAIM
   39S   09W   02
   42-12-20N   123-43-08W
   SAXONITE-DUNITE   81 03
LOW GAP CLAIM
   40S   05W   34
   42-02-27N   123-16-44W
   QUARTZITE   81 03
LOW RIDGE CLAIM
   37S   09W   16
   42-20-51N   123-46-33W
   SERPENTINE   81 03
LOWER GRAVE CREEK PLACERS
   34S   07W   02
   42-38-32N   123-30-07W
   AU   STREAM AND BENCH GRAVELS
LUCKY ANTLER
   36S   08W   12   "NE,SW"
   42-27-12N   123-35-29W
   AU   META-ANDESITE SERPENTINE 81 03
LUCKY HUNCH CLAIM
   37S   09W   33
   42-18-06N   123-45-31W
   CR   81 03
LUCKY L. & R.
   35S   09W   35
   42-29-04N   123-43-43W
   CR   DUNITE SAXONITE   81 03
LUCKY PAT CLAIM
   37S   09W   29
   42-19-26N   123-46-47W
   CR   DUNITE   81 02
LUCKY PENNY
   41S   06W   16   NW
   42-00-32N   123-25-17W
   METAGABBRO 81 03
LUCKY QUEEN MINE
   34S   05W   30   "SW,SE"
   42-34-49N   123-20-04W
   AU   81 03
LUCKY SHOT
   34S   08W   12
   42-37-52N   123-35-37W
   AU AG PORPHYRITIC DACITE 81 03

LUCKY SPOT
    36S   07W   29   "NE,NW"
    42-24-60N   123-33-16W   ULTRAMAFICS
    81 03

LUCKY SPOT
    39S   09W   03   N
    42-12-45N   123-44-46W   ULTRAMAFICS
    81 03
LUCKY STAR
    37S   09W   21
    42-20-34N   123-46-22W
    CR   SERPENTINE   81 03
LUCKY STRIKE
    39S   08W   18
    42-10-29N   123-40-47W
    CR PGM   SAXONITE   81 03
LUCKY STRIKE
    41S   07W   05   SW
    42-01-37N   123-33-18W
    AU   METASEDIMENTS   81 03
LUCKY STRIKE
    41S   06W   07   NW
    42-01-03N   123-27-38W
    AU CU GREENSTONE 81 03

LUCKY STRIKE - CASTLE SPRINGS
    37S   09W   30
    42-19-16N   123-47-55W
    CR   DUNITE   81 03
LUETHYE MINE
    40S   07W   26   "NE,NE"
    42-03-57N   123-29-09W
    METAVOLCANICS   81 03
LYONS (WOOD) PROSPECT
    34S   06W   23   NE
    42-36-16N   123-22-08W
    "DUNITE ON OLIVINE SAXONITE, ALTERED TO
    SERPENTINE"
LYTTLE MINE
    41S   08W   01   SW
    42-01-30N   123-35-48W
    CU   GREENSTONE 81 03
MACABEE MINE
    33S   05W   20   "SW,SE"
    42-40-56N   123-18-58W
    AU   GREENSTONE SLATE   81 03

MACFARLANE BRICK PLANT
        37S   08W   24   SW
        42-20-07N   123-35-09W
MADISON GULCH 1 & 2
        39S   09W   02   "S,NW"
        42-12-27N   123-43-56W
                SERPENTINE   81 03
MAMMOUTH CHROME
        33S   05W   28
        42-40-27N   123-17-45W
MANGANESE PROSPECT
        41S   05W   02   "W,W"
        42-01-30N   123-16-05W
        QUARTZITE   81 03
MANSFIELD PROSPECT
        39S   07W   27   "SE,SW"
        42-08-33N   123-30-56W
        SERPENTINE
MARVIN MINE
        34S   08W   22
        42-36-20N   123-37-58W   PYROXENITE
METAGABBRO AMPHIBOLITE
MARY WALKER
        36S   09W   27
        42-24-59N   123-45-06W
        CR   SERPENTINE-DUNITE   81 03
MAX CAMPBELL
        36S   07W   05
        42-28-18N   123-33-18W
        QUARTZITE   81 03
MAY QUEEN MINE
        36S   05W   26   "W,W"
        42-24-45N   123-16-00W
        AU   "HARD, DENSE GREENSTONE" 81 03
MAY QUEEN PLACER
        40S   07W   28   "SW,SW"
        42-03-11N   123-31-19W
        SERPENTINE AND PORPHYRY BEDROCK
MAYFLOWER GROUP
        34S   08W   27   "S,NW"
        42-35-15N   123-38-08W
        AMPHIBOLITE
MAYNARD – FRENCH
        33S   05W   26   "NE,SE"
        42-40-22N   123-14-56W   SLATE
MC GRATH
        38S   09W   26   "SW,SW"
        42-13-45N   123-44-10W
        AU AG   81 03

MC NAIR FLAT PLACER
    34S   07W   04   "NW,NW"
    42-39-06N   123-32-30W
    AU   BENCH GRAVEL   81 05
MC PHERSON PROSPECT
    39S   09W   07
    42-11-25N   123-47-49W
    GREENSTONE AND QUARTZITE.   81 04
MEATHEAD
    33S   05W   26   "NW,NE"
    42-40-46N   123-15-17W
    SLATE 81 03
MEL BARLOW
    39S   08W   27
    42-09-47N   123-37-50W
    SAND & GRAVEL
MERRILL PLACER
    40S   08W   27
    42-03-31N   123-37-53W
MICHIGAN MINE
    37S   05W   16   "W,SW"
    42-20-59N   123-18-19W
    AU   GREENSTONE 81 03
MIDDLEMAN PROSPECT
    40S   07W   24
    42-04-30N   123-28-50W
    SLATE SHALE 81 03
MIDNIGHT CLAIM
    37S   09W   21
    42-20-28N   123-46-16W
    CR   SERPENTINE   81 03
MIGHTY JOE
    39S   09W   13
    42-10-26N   123-43-00W
    CR   SERPENTINE   81 03
MILL CREEK PROSPECT
    34S   08W   32   NW
    42-34-36N   123-40-19W   AMPHIBOLITE
MILLERS DREAM PROSPECT
    37S   09W   32
    42-18-23N   123-47-21W
    CR   DUNITE   81 03
MOCKINGBIRD
    37S   09W   29
    42-19-02N   123-46-56W
    CR   SERPENTINE   81 03

MOHAWK CLAIM
     38S   09W   05
     42-17-35N   123-47-18W
     CR   SERPENTINE   81 03
MOLLY GROUP
     40S   09W   29   E
     42-03-35N   123-46-57W   ULTRAMAFICS
     81 03
MOLLY HILL MINE
     33S   08W   26   "NE,SE"
     42-40-19N   123-36-06W   PORPHYRITIC
META-ANDESITE   81 03
MOLYBDENUM OCCURRENCE
     35S   09W   01
     42-33-27N   123-42-12W
     QUARTZ DIORITE
MOLYBDENUM OCCURRENCE
     35S   06W   08   NW
     42-32-45N   123-26-29W
     PEGMATITE

MOLYBDENUM OCCURRENCE
     35S   06W   31
     42-29-22N   123-26-48W
     QUARTZ DIORITE
MOODY
     38S   09W   34   W
     42-13-15N   123-44-49W   GREENSTONE
     81 03
MOONBEAM MINE
     33S   08W   33   "W,NW"
     42-39-49N   123-38-50W
     AU   HORNBLENDE GABBRO   81 03
MOORE CHROMITE
     33S   05W   28
     42-40-47N   123-17-32W
     CR   SAXONITE   81 03
MOORE LANE
     37S   05W   21
     42-20-03N   123-17-36W
     SAND & GRAVEL

MORNING MINE
     33S   05W   20   NW
     42-41-25N   123-19-20W
MOUNT PEAVINE
     34S   08W   21
     42-36-08N   123-38-31W

MOUNT PITT
    34S   05W   36   "SE,SE"
    42-33-55N   123-13-50W
    AU   METABASALT ARGILLITE   81 03
MOUNT REUBEN
    33S   08W   12
    42-43-20N   123-35-38W
    AU AG U   GREENSTONE 81 03
MOUNTAIN LION PROSPECT
    37S   05W   25
    42-19-35N   123-14-40W
    AU AG GREENSTONE 81 03
MOUNTAIN TOP QUARRY
    36S   07W   36
    42-23-30N   123-27-55W   STONE
MOUNTAIN TOP QUARRY
    37S   07W   01   SW
    42-22-45N   123-28-50W   STONE
MOUNTAIN TREASURE MINING CO.
    34S   05W   35
    42-33-58N   123-15-34W
    GREENSTONE AND SERPENTINE   81 04
MOUNTAIN VIEW (MITCHELL)
    40S   05W   23   "E,E"
    42-04-29N   123-15-01W
    AU   METAVOLCANICS   81 03
MUCK LIMESTONE
    37S   06W   30   "SW,NW"
    42-19-32N   123-27-45W
    STONE(LIMESTONE)   LIMESTONE   81 04
MULE SHOE CLAIM
    37S   09W   21   "SEE,NW"
    42-20-24N   123-46-04W
    CR   DUNITE   81 03
MURPHY PLACER
    37S   05W   25
    42-19-09N   123-14-34W
MURRAY MINE
    36S   07W   07   SE
    42-27-04N   123-33-59W
    SLATE 81 03
NEIL
    38S   09W   34
    42-13-05N   123-45-13W   GREENSTONE
    81 02
NESBIT GROUP
    34S   08W   34   "NE,SW"
    42-34-14N   123-38-01W
    AU   81 03

NETTIE CHROME CLAIM
    39S   09W   25
    42-08-57N   123-42-07W   ULTRAMAFICS
    81 03
NEW DEAL CLAIM
    39S   09W   15   SW
    42-21-06N   123-45-03W
          SERPENTINE   81 03
NEW ELDORADO
    34S   05W   17
    42-37-14N   123-19-14W
NICHOLS PLACE
    37S   09W   21   SW
    42-20-05N   123-46-15W
NICHOLS PROSPECT
    37S   09W   28   "SW,NE"
    42-19-29N   123-45-53W
          SERPENTINIZED DUNITE   81 03
NICKEL RIDGE GROUP
    40S   09W   31
    42-02-21N   123-47-48W
    CR   81 03
NICKEL RIDGE LATERITE
    40S   09W   31   E
    42-02-46N   123-47-51W
    "SERPENTINITE, PERIDOTITE" 81 03
NOBLE BAR
    38S   05W   35   "SE,SE"
    42-18-20N   123-15-11W
          SAND & GRAVEL   93 04
NORTHERN CALIFORNIA DREDGING CO.
    34S   05W   25
    42-35-00N   123-13-45W   AU
NUMBER 8 GULCH
    40S   07W   22
    42-04-54N   123-30-49W   AU
OAK MINE
    35S   05W   04
    42-33-02N   123-18-25W
    AU AG METABASALT 81 03
OCCURRENCE
    35S   08W   08   "NW,SE"
    42-32-29N   123-40-09W
OLD CASEY
    38S   09W   06
    42-17-55N   123-48-11W   CR   81 03

OLD CHANNEL
      34S   08W   35   NE
      42-34-26N   123-36-31W
      AU   TERRACE GRAVEL   81 05
OLD CHANNEL MINE
      35S   08W   03   NE
      42-33-51N   123-37-39W   STONE
OLD CHANNEL PLACERS
      35S   08W   10   "N,NW"
      42-32-58N   123-38-02W
      AU   SEMI-CONSOLIDATED GRAVELS
OLD CROW
      33S   05W   27
      42-40-26N   123-17-12W
OLD GLORY MINE
      36S   10W   12   NE
      42-27-36N   123-49-16W
      AU AG HORNBLENDE DIORITE   81 03

OLD RAY
      38S   08W   19
      42-14-36N   123-40-44W   AU
OLD SMOKEY
      34S   05W   26   CENTER
      42-35-04N   123-15-30W
      SERPENTINE 81 03
OLD SMOKEY
      39S   09W   14   SW
      42-10-22N   123-43-46W   ULTRAMAFICS
      81 03
ONION FALLS COPPER
      36S   08W   02   "S,NE"
      42-28-15N   123-36-20W
      CU   METAVOLCANICS   81 02
ONION MOUNTAIN NICKEL
      36S   07W   06   SW
      42-27-54N   123-34-36W   SERPENTINITE
      81 03
ONION MOUNTAIN QUARRY
      36S   08W   33
      42-23-54N   123-39-00W   STONE
ONION MTN QUARRY
      36S   05W   17
      42-26-10N   123-19-23W   STONE
OREG. LIME PRODUCTS MINE
      38S   05W   15   SW
      42-15-38N   123-16-49W   STONE(LIMESTONE)
      81 03

OREGON BEAUTY
    38S   05W   15
    42-15-37N   123-17-10W
    METAVOLCANICS   81 03
OREGON BONANZA MINE
    38S   05W   16   SE
    42-15-41N   123-17-27W
    AU AG METASEDIMENTS   81 03
OREGON BONANZA MINE (?)
    38S   05W   10   "NW,SW"
    42-16-50N   123-17-10W
    METAVOLCANICS   81 03
OREGON CALCITE CORPORATION
    40S   06W   16
    42-05-24N   123-24-34W
OREGON CAVES NAT. MONUMENT
    40S   06W   09   SE
    42-05-54N   123-24-23W   81 03
OREGON CHROME MINE
    39S   09W   21   "N,N"
    42-20-38N   123-46-08W
    CR   SERPENTINIZED DUNITE   81 03
ORIOLE
    34S   08W   26   NE
    42-35-27N   123-36-33W
    AU AG GREENSTONE RHYODACITE DACITE
    PORPHYRY   81 03
ORO FINO
    35S   05W   03   "SE,SE"
    42-33-00N   123-16-11W
    AU   GREENSTONE 81 03
ORO GRANDE
    33S   05W   28   "SE,SE"
    42-40-02N   123-17-20W
    AU   SERPENTINE 81 03

OSCAR CREEK PLACER
    37S   05W   22
    42-20-13N   123-16-45W
    CHANNEL AND BENCH GRAVELS   81 05
OW YUEN CLAIMS
    40S   06W   05   "NW,NW"
    42-07-30N   123-26-22W
    QUARTZITE   81 03
OWEN CLAIM
    41S   07W   07
    42-00-46N   123-33-55W
    CHERT 81 03

OWENS
      41S   08W   11
      42-01-16N   123-36-45W
      CR   SERPENTINE   81 03

P.D.Q. CLAIM
      36S   09W   02
      42-27-54N   123-44-11W
      CR   DUNITE AND SAXONITE   81 03

PANTHER CREEK MINING CO.
      37S   09W   06
      42-22-42N   123-47-46W   AU

PARADISE NO. 1
      39S   06W   04
      42-12-52N   123-24-47W
      CR   SERPENTINE   81 04

PARADISE NO.2
      38S   06W   33
      42-13-11N   123-24-49W
      CR   SERPENTINE   81 04

PARKER RIDGE
      40S   09W   11   "S,S"
      42-05-47N   123-43-36W

PAYDAY
      39S   08W   07   NW
      42-11-41N   123-41-37W   ULTRAMAFICS
      81 04

PEACH
      37S   09W   29
      42-19-29N   123-47-16W   CR

PEAVINE
      34S   08W   27   "SE,SE"
      42-35-01N   123-37-35W
      LATERITES

PEGI
      33S   05W   27   NW
      42-40-37N   123-16-51W
      SERPENTINE   81 04

PENNINGTON BUTTE
      38S   05W   08
      42-17-10N   123-19-21W
      CR   81 04

PHILLIP
      38S   09W   23   NW
      42-15-20N   123-43-58W   GREENSTONE

PICKETT CREEK
      35S   07W   27   NE
      42-30-02N   123-30-25W   STONE

PICKETT CREEK
    35S   07W   33
    42-28-45N   123-31-45W
    HG   SANDSTONE AND SHALE   81 02

PLACER
    38S   09W   05
    42-17-37N   123-47-50W
PLACER MINE
    34S   05W   24   "NW,NE"
    42-36-25N   123-14-20W
PLACER MINER CO.
    37S   09W   06
    42-22-35N   123-48-11W   AU
PLANT ON J ST.
    37S   05W   21
    42-20-25N   123-17-57W   SAND & GRAVEL
PLATERICA MINE
    40S   08W   33   "NE,NE"
    42-03-13N   123-38-08W
    AU   CONGLOMERATE
PONY SHOE CHROME
    41S   07W   14
    42-00-25N   123-30-22W
    CR   SERPENTINE   81 04
POORMAN NORTH MINE
    33S   08W   23   "E,NW"
    42-41-29N   123-36-46W
    AU   GABBRO   81 04
PORCUPINE MINE
    37S   05W   23   W
    42-20-26N   123-15-44W
    METAVOLCANICS   81 04
PORPHYRY GROUP
    33S   05W   28
    42-40-36N   123-18-12W
PORTLAND GROUP
    40S   07W   03   CENTER
    42-07-05N   123-30-43W
    AU AG ARGILLITE   81 04
POSTEM
    35S   08W   36
    42-29-02N   123-35-31W
    ALTERED SERPENTINITE   81 02
POWELL CREEK
    38S   05W   16
    42-15-56N   123-17-51W   FE

PROSPECT
    34S   08W   25   "NE,NW"
    42-35-31N   123-35-37W
PROSPECT
    35S   08W   03   SW
    42-33-11N   123-38-02W
PROSPECT
    40S   07W   14   "E,SE"
    42-05-12N   123-29-08W   GREENSTONE
    81 04
PUGH SOAPSTONE
    38S   05W   17   "SE,SW"
    42-15-43N   123-19-08W   TALC(SOAPSTONE)
    SERPENTINE

PYLES BAR
    33S   08W   32   "SE,SW"
    42-39-16N   123-40-24W
PYX MINE
    33S   08W   26   "NW,NE"
    42-40-51N   123-36-35W
    AU   PORPHYRITIC META-ANDESITE   81 04
QUARTETTE
    38S   05W   29
    42-14-14N   123-19-05W   AU
QUEEN MINE
    36S   07W   30   "CENTER,N"
    42-24-54N   123-34-20W   GREENSTONE
    81 04
QUEEN OF BRONZE MINE
    40S   08W   36
    42-02-59N   123-35-55W
    CU AU AGGREENSTONE (METABASALT AND
    METAGABBRO)   81 04
R R R METAL
    39S   06W   27
    42-09-12N   123-23-57W
RAINBOW
    33S   07W   20   SW
    42-41-10N   123-33-24W   GREENSTONE
(SERICITE SCHIST)   81 04
RAINBOW
    33S   04W   13
    42-42-08N   123-13-51W
    CR   SERPENTINE   81 04
RAINBOW MINE
    40S   07W   01   "S,S"
    42-06-34N   123-28-54W
    AU AG GREENSTONE 81 04
242

RAMSEY MINE

    36S   08W   24   "SW,SW"

    42-25-12N   123-35-49W   AU   GREENSTONE

    81 04

RAMSEY MINE

    41S   07W   09   NE

    42-01-18N   123-31-38W

    GREENSTONE 81 04

RAND PLACER

    34S   07W   19   "W,NW"

    42-36-18N   123-34-45W

    AU   BENCH GRAVELS   81 05

RAYBOY

    38S   07W   15

    42-15-51N   123-30-44W

RED DEVIL

    36S   08W   12   NE

    42-27-30N   123-35-19W

    GREENSTONE SERPENTINE

RED DOG

    37S   09W   03

    42-23-10N   123-45-05W   AU

RED DOG (VAN GWYN GROUP)

    36S   09W   34   W

    42-23-47N   123-44-38W   CR

RED ELEPHANT MINE

    34S   08W   19   SW

    42-35-55N   123-41-41W

    QUARTZ

RED JACKET MINE

    35S   05W   34   NE

    42-29-18N   123-16-13W

    AU AG GREENSTONE 81 04

RED LEDGE

    36S   08W   01   "CENTER,S,S"

    42-27-44N   123-35-29W

    ANDESITE TUFF AND FLOWS (CHLORITIZED) 81 02

REED

    33S   05W   28

    42-40-33N   123-18-09W   AU

REED

    33S   05W   28   "SW,NW"

    42-40-32N   123-18-05W

    AU   GREENSTONE GABBRO SERPENTINE

REEVES CREEK QUARRY
        38S   08W   26   NW
        42-14-17N   123-36-49W
        STONE
RENO
        33S   08W   34   NW
        42-39-53N   123-38-16W
        AU AG GREENSTONE 81 04
REVELL
        38S   08W   07   "NE,SW"
        42-16-44N   123-41-24W
        AU   RIVER GRAVEL & HILLSIDE SLUMP
REYNOLD'S COPPER MINE
        40S   09W   27
        42-03-44N   123-45-24W
        SERPENTINE 81 04
RHEA
        37S   08W   23
        42-18-17N   123-38-57W
RICHMOND GROUP
        34S   08W   23   NE
        42-36-24N   123-36-23W   GREENSTONE
        81 04
RISING STAR MINE
        38S   05W   21   N
        42-15-27N   123-17-27W   AU   HORNBLENDE
SCHIST 81 04
RIVERSIDE
        37S   09W   33
        42-18-42N   123-45-42W
ROCK CREEK
        41S   09W   03
        42-01-51N   123-44-45W
        CR   81 04
ROCKY GULCH
        34S   08W   22
        42-36-22N   123-37-35W
        AU   BANDED QUARTZITE 81 04
ROCKY GULCH PLACER
        34S   08W   25   S
        42-34-53N   123-35-32W
RORRICK
        40S   07W   02
        42-06-49N   123-30-10W   AU
ROSE CITY MINE
        36S   09W   24   "NW,NW"
        42-25-55N   123-43-00W
        PERIDOTITE   81 02

ROSE CLAIM
     37S   09W   28
     42-19-04N   123-46-16W
     DUNITE   81 03
ROSE QUARTZ
     38S   05W   29
     42-14-05N   123-19-00W   AU
ROSEBURG AND FIDELITY GROUP
     39S   08W   07
     42-11-28N   123-41-19W
     AU   GREENSTONE SERPENTINE   81 04
ROUGH AND READY BENCH
     40S   10W   01   E
     42-06-60N   123-49-05W
     NI   LATERITES   81 04

ROUGH AND READY CREEK
     40S   09W   14
     42-05-22N   123-45-05W   LATERITES
ROUGH AND READY CREEK
     40S   09W   16   SW
     42-05-07N   123-45-37W
     NI   LATERITES   81 04
ROUGH AND READY OUTWASH
     40S   09W   13
     42-05-13N   123-42-11W
     NI   LATERITES   81 04
ROUGH AND READY RIDGE
     40S   09W   22
     42-04-29N   123-45-56W
     NI   LATERITES   81 04
SAD SACK
     36S   09W   23
     42-25-43N   123-43-11W
     CR   81 04

SADDLE
     38S   10W   02
     42-17-08N   123-48-58W   SAXONITE
SADDLE CHROME
     38S   09W   06
     42-17-59N   123-48-26W
     CR   81 04
SAINT PATRICK MINE
     33S   08W   35   "NW,SE"
     42-39-29N   123-36-38W
     AU   GREENSTONE (ANDESINE - HORNBLENDE
     SCHIST)

SALLY ANN
    36S   08W   36
    42-23-40N   123-35-27W
SALLY ANN CLAIM
    37S   09W   16
    42-21-09N   123-46-31W
    CR   81 04

SALT ROCK
    36S   07W   06
    42-28-16N   123-33-47W   CR
SATURDAY ANNE
    37S   09W   09
    42-22-09N   123-46-08W
    CR   DUNITE
SCANDINAVIAN-AMERICAN PLACER
    34S   08W   12
    42-37-35N   123-35-45W   AU
SCHOLEY GULCH
    33S   05W   28   N
    42-40-36N   123-17-48W   SERPENTINE
SCOTTY
    33S   08W   23   "S,NW"
    42-41-29N   123-36-57W   METAGABBRO
    81 03
SEVEN-THIRTY MINE (JUANITA MINE)
    34S   08W   14   "N,NW"
    42-37-22N   123-37-03W
    AU   GREENSTONE
SEXTON MOUNTAIN
    34S   06W   24
    42-36-12N   123-21-08W
    CR   SERPENTINE   81 04
SHADY COVE
    36S   09W   11
    42-27-27N   123-43-29W
    CR   PERIDOTITE   81 04
SHERMAN PLACER
    37S   09W   32
    42-18-50N   123-47-00W
    AU
SHOT
    33S   05W   33   NW
    42-39-53N   123-18-10W
    AU AG METAGABBRO GREENSTONE
SILENT FRIEND MINE
    33S   05W   15   "W,SE"
    42-41-58N   123-16-36W
    AU AG GREENSTONE 81 04
246

SILVER NUGGET PLACER
    39S   09W   12   "SW,SW"
    42-11-11N   123-42-52W
SILVER TIP
    38S   06W   25
    42-14-22N   123-21-42W   CR
SILVERTIP CHROME
    38S   06W   36
    42-13-20N   123-21-11W
SIXMILE COPPER
    37S   09W   35   SW
    42-18-17N   123-43-49W
    GREENSTONE/SERPENTINITE 81 04
SIXMILE CREEK PLACERS
    38S   09W   02   N
    42-17-50N   123-43-45W
    AU   TERRACE AND CHANNEL GRAVELS
SKINNER PROSPECT
    33S   08W   27   "SW,SW"
    42-40-03N   123-38-22W
    FINE GRAINED GREENSTONE 81 04
SLUG BAR/GYPSEY QUEEN
    40S   07W   15
    42-05-11N   123-31-05W   AU
SLUTER MINE
    33S   05W   11   "S,NW"
    42-43-03N   123-15-41W
    AU   GREENSTONE 81 04
SNOOK
    34S   05W   19
    42-36-02N   123-20-07W
    SERPENTINE 81 04
SNOW BIRD
    38S   05W   20   NE
    42-15-24N   123-18-53W
    AU   METASEDIMENTS   81 04
SORDY CLAIM
    34S   08W   23   "W,NE"
    42-36-17N   123-36-34W
    GREENSTONE AND SERPENTINE
SORDY PLACER
    34S   08W   36   W
    42-34-19N   123-35-51W
SOUTH FLANK OF ROUGH MOUNTAIN
    40S   09W   08
    42-06-17N   123-47-58W
    LATERITES

SOWELL CHROMITE
> 40S   08W   30
> 42-03-36N   123-41-20W

SPEAKER PLACER
> 33S   05W   08   SE
> 42-42-43N   123-18-34W
> AU   CHANNEL AND TERRACE GRAVELS

SPECTOR
> 35S   07W   32   S
> 42-28-48N   123-32-31W
> GREENSTONE AND SERPENTINE   81 04

SPENCE MINE
> 40S   07W   19
> 42-04-28N   123-34-16W
> GREENSTONE 81 04

SPOKANE GROUP
> 34S   08W   34   "NE,NE"
> 42-34-38N   123-37-26W
> SERPENTINE AMPHIBOLITE   81 04

SPOTTED FAWN
> 33S   05W   22   E
> 42-41-25N   123-16-20W
> AU AG   GREENSTONE 81 04

SPRADLINGE
> 35S   08W   16   "SW,NE"
> 42-31-52N   123-24-22W

SQUAW CREEK CHROME
> 38S   08W   04
> 42-17-26N   123-39-07W   CR

ST. PAUL COPPER
> 33S   05W   35   "NW,NW"
> 42-39-55N   123-15-53W
> METAVOLCANICS   81 02

STANDARD GROUP
> 34S   08W   24
> 42-35-40N   123-35-00W   AU

STAR MINE
> 34S   05W   08   E
> 42-37-49N   123-18-42W
> GREENSTONE 81 04

STATE CLAIM
> 40S   06W   30   "N,NE"
> 42-04-03N   123-26-54W   GREENSTONE
> 81 04

STEVENS –MILLER
> 38S   05W   07
> 42-16-51N   123-20-16W
> CR   SERPENTINE   81 04

STONER BAR
       36S   06W   20
       42-25-47N   123-26-13W
       SAND & GRAVEL
STORE GULCH CLAIM
       38S   09W   03
       42-18-04N   123-44-44W
       CR   SERPENTINE   81 04

STOVEPIPE MINE
       34S   06W   17   NW
       42-37-12N   123-26-20W
       AU   TERRACE GRAVEL   81 05
STRATON CREEK
       35S   07W   05
       42-33-30N   123-33-20W   AU
STRATTON CREEK PLACER
       35S   07W   04   "NW,SW"
       42-33-22N   123-32-29W
       AU   TERRACE GRAVELS   81 05
STRENUOUS TEDDY
       35S   08W   05
       42-33-37N   123-40-06W
       AU AG AMPHIBOLITE 81 04
SUCKER CREEK PLACERS
       40S   06W   01   CENTER
       42-07-37N   123-28-01W
       AU   81 04

SUCKER CREEK SAND BAR
       39S   06W   31
       42-08-12N   123-27-49W   AU
SUGAR LOAF
       33S   07W   19
       42-41-24N   123-34-14W   AU
SUGAR PINE
       35S   08W   03   "SW,NW"
       42-33-37N   123-38-21W
       AU AG AMPHIBOLITE 81 04
SUGAR PINE PLACER
       35S   08W   05   "SE,NW"
       42-33-32N   123-40-23W
SULFIDE NO. 1
       38S   09W   23   "CENTER,SW"
       42-14-51N   123-43-59W
       METAVOLCANICS   81 03

SULFIDE OCCURRENCE
        34S   06W   22   "SE,SE"
        42-35-39N   123-23-15W
        VOLCANIC SEDIMENTS   81 04
SUMMIT GROUP
        38S   09W   19   CENTER
        42-14-51N   123-48-24W
        AU   GREENSTONE (ANDESITE PORPHYRY)
SUNNY MAYDAY
        38S   05W   19   W
        42-15-03N   123-19-49W
        CHERT 81 04
SUNSET
        40S   07W   22
        42-04-09N   123-31-17W   AU
SUNSET
        41S   05W   16   SW
        42-00-21N   123-17-56W
SUNSET MINE
        33S   05W   14   "NE,NW"
        42-42-27N   123-15-36W
SURPRISE MINING COMPANY
        41S   07W   03
        42-01-39N   123-30-48W

SWASTIKA MINING CO.
        34S   05W   32
        42-34-22N   123-19-07W
        LOW TERRACE GRAVEL   77 05
SWEDE BASIN PROSPECT
        37S   09W   01   NW
        42-23-01N   123-41-20W
        FELSITIC AND AMYGDALOIDAL GREENSTONES
SWEDE CREEK OCCURRENCE
        36S   09W   34   SE
        42-23-40N   123-44-10W   ULTRAMAFICS
        81 04
SWEDE QUARRY
        36S   05W   17
        42-26-10N   123-19-23W   STONE
TANNEN PLACER
        41S   06W   04
        42-02-05N   123-25-00W   AU
TAYLOR CAMP QUARRY
        36S   05W   17
        42-26-10N   123-19-23W   STONE
TENNESSEE BAR
        33S   09W   23
        42-41-01N   123-34-07W   AU

TENNESSEE CHROMITE
    39S   09W   11
    42-11-45N   123-43-12W
    CR   PERIDOTITE   81 04
TENNESSEE PASS
    39S   08W   07
    42-11-35N   123-42-18W
    CR PT   SERPENTINE   81 04
TEXAS OREGON POWER & PLACER
    35S   07W   06   "SE,SE"
    42-33-02N   123-33-48W
THOMPSON GROUP
    33S   05W   26
    42-40-45N   123-15-53W
    CR   SERPENTINE   81 04
THREE L CLAIM
    35S   09W   35
    42-29-02N   123-44-15W
    SERPENTINE   81 04
THREE LODES MINING CO.
    34S   08W   34
    42-34-00N   123-38-00W
THREE L'S PLACER (YOKUM)
    34S   07W   03   "E,E"
    42-38-40N   123-30-22W
    AU   CHANNEL AND TERRACE GRAVELS
TIBBETT SPRINGS MINE
    33S   08W   12   "W,NE"
    42-43-19N   123-35-24W
    GREENSTONE (ANDESITE PORPHYRY AND TUFF)

TIP TOP
    40S   07W   11
    42-06-31N   123-29-15W
    AU AG METASEDIMENTS   81 04

"TOMLINSON, GATES & THOMAS CLAIMS"
    41S   06W   05   SE
    42-01-37N   123-25-33W   GREENSTONE
    81 04
TOOTS MINE
    38S   08W   19   NE
    42-15-15N   123-41-00W
    CR   SERPENTINE   81 04
TRADE DOLLAR MINE
    33S   08W   23   NE
    42-41-29N   123-36-22W
    AU   GREENSTONE 81 04

TREASURY GROUP
    34S   08W   20   "SE,SE"
    42-35-41N   123-39-46W
    GABBROIC AMPHIBOLITE   81 04
TURNER-ALBRIGHT
    41S   09W   09   "SE,NE"
    42-00-40N   123-45-42W
    CU AU METABASALT CHERT 81 02
TURVEY
    38S   05W   29   N
    42-14-20N   123-19-01W   MARBLE
TWELVE O'CLOCK MINE
    35S   05W   04   NE
    42-33-33N   123-17-39W   GREENSTONE
TWIN CEDARS CLAIMS
    38S   09W   06
    42-17-31N   123-48-30W
    CR   DUNITE   81 04
TYEE BAR
    33S   08W   33   "NW,SW"
    42-39-26N   123-39-23W
    BEDROCK IS ARGILLITE.
U. S. PLACER
    39S   07W   29
    42-08-35N   123-33-00W
UPPER FALL CREEK LODE & PLACER
    38S   09W   16
    42-16-11N   123-45-58W

UPPER HAMMON
    37S   09W   21
    42-19-37N   123-45-43W
UPPER PINE CREEK
    36S   10W   23   SW
    42-25-17N   123-50-50W
    AMPHIBOLITE GNEISS 81 03
VALEN PROSPECT
    40S   07W   31
    42-02-37N   123-33-57W
    CR   SERPENTINIZED PERIDOTITE 81 04
VICTOR
    33S   05W   20   "NW,SW"
    42-41-11N   123-19-27W
    METASEDIMENTS   81 04

VICTOR
    34S   08W   32
    42-33-56N   123-39-47W
    AU AG AMPHIBOLITE 81 04

VINDICATOR PLACER
    33S   07W   34   "NE,SW"
    42-39-29N   123-30-51W
    AU   TERRACE GRAVELS   81 05
VIOLET
    36S   09W   14
    42-26-33N   123-43-40W
    CR   DUNITE   81 04
W. C. HURST CLAY
    36S   07W   30
    42-24-40N   123-34-15W
WALDO COPPER MINE
    40S   08W   36   "SW,"
    42-02-26N   123-35-53W
    CU AU GREENSTONE SERPENTINE
WALDO HILL MINE & MILL
    41S   08W   04
    42-01-47N   123-38-48W
WASHINGTON BRICK & LIME CO.
    38S   05W   22
    42-15-17N   123-16-34W   STONE(LIMESTONE)

WATERS CREEK ALUMINUM PROSPECT
    36S   07W   32
    42-24-00N   123-33-20W
WEBSTER ROAD
    36S   06W   24   NE
    42-25-44N   123-21-18W
    SAND & GRAVEL(TOPSOIL)
WEST FORK OF ILLINOIS RIVER
    40S   09W   27   SE
    42-03-10N   123-44-52W
    LATERITES
WESTERN NO. 1 CLAIM
    35S   09W   35
    42-28-47N   123-44-03W
    DUNITE
WESTERN NO. 2
    35S   09W   35
    42-28-38N   123-44-09W
    DUNITE   81 04
WHISKEY SPRINGS MINE
    33S   08W   28   "S,NE"
    42-40-26N   123-38-46W
    GREENSTONE 81 04
WHITE PINE MINE
    38S   09W   14   "S,NW"
    42-15-55N   123-44-00W
    GREENSTONE 81 04

WHITE ROCK II QUARRY
    38S   08W   35
    42-13-20N   123-36-30W   STONE
WHITE WATER LODE CLAIM
    38S   09W   05
    42-17-19N   123-46-52W
    CR   SERPENTINE   81 04
WHITE WONDER
    35S   06W   01   NW
    42-33-34N   123-21-34W
    PERIDOTITE   81 04
WHITENECK GROUP
    33S   08W   27   "NW,SE"
    42-40-17N   123-37-43W
    GABBRO AND RECRYSTALLIZED GREENSTONE
        81 04
WILD CAT CLAIM
    40S   07W   32
    42-02-32N   123-33-27W
    CR   SERPENTINE   81 04
WILD DEER PROSPECT
    38S   05W   17
    42-16-12N   123-18-59W
    MN   QUARTZITE   81 04
WILLIAMS AND ADYLOTT
    38S   09W   10   E
    42-16-43N   123-44-26W
    AU AG GREENSTONE (TUFF AND TUFF BRECCIAS)
        81 04
WILLIAMS PIT
    38S   05W   26   NE
    42-14-02N   123-15-39W   SAND & GRAVEL

WINDY GAP MINE
    33S   08W   11   "SE,NE"
    42-43-09N   123-36-08W   GREENSTONE
81 04
WINTERS PROSPECT
    39S   09W   08
    42-11-42N   123-47-16W   GREENSTONE
& SLATES   81 04
WOLF CREEK LOGGING AND MINING COMPANY   33S
    05W   10
    42-43-02N   123-17-13W
WOLF CREEK PLACERS
    33S   06W   05
    42-41-08N   123-22-58W   AU

WOLFE PROSPECT
       35S   07W   16
       42-31-42N   123-31-55W
WOOD PROSPECT
       34S   06W   23
       42-36-13N   123-22-12W
       DUNITE

WOODCOCK CREEK
       39S   09W   36
       42-07-56N   123-42-30W
WOODCOCK MOUNTAIN
       39S   09W   25
       42-08-21N   123-42-07W
       LATERITES   79 03
YANKEE CHIEF PLACER
       35S   08W   16   NE
       42-32-22N   123-38-41W
       AU   CHANNEL GRAVELS   81 05
YANKEE SILVER LODE
       34S   08W   25   "NW,SW"
       42-35-11N   123-36-07W
       SILTSTONE/TUFF   81 04
YEAGER MINE
       40S   07W   12   "NW,SE"
       42-06-29N   123-28-38W

YELLOW JACKET CLAIM
       36S   09W   22
       42-25-51N   123-44-18W   CR   DUNITE
YELLOWHORN
       34S   05W   04   SW
       42-38-23N   123-18-32W
       AU AG GREENSTONE 81 04
YOUNG PROSPECT
       33S   05W   06
       42-43-58N   123-20-13W
       SCHISTOSE GREENSTONE   81 02
YOUNG'S MINE
       38S   09W   06
       42-17-57N   123-48-46W   CR   DUNITE
YOUNGS PLACER
       36S   08W   18
       42-26-13N   123-41-51W   AU
ZIG-ZAG PROSPECT
       33S   05W   25   SW
       42-40-09N   123-14-41W   SILTSTONE

Jackson County

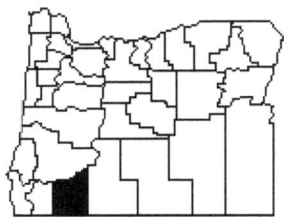

NAME

TOWNSHIP   RANGE   SECTION   SECTION FRACTION

NORTH LATITUDE   WEST LONGITUDE

MINED MATERIAL

Last Update of Claim or Quarry if I was able to get it

140 PIT
       36S   01E   08
       42-26-53N   122-43-39W   STONE

140 ROCK QUARRY
       36S   01W   26   SE
       42-24-32N   122-46-44W   STONE

ACE OF HEARTS
       38S   04W   12
       42-16-47N   123-07-13W   AU

AFTERTHOUGHT
       38S   04W   27   SE
       42-14-15N   123-09-32W
       AU   "ARGILLITE, SANDSTONE"   80 12

AL SARENA
       31S   02E   29
       42-50-43N   122-36-19W
       AU AG PB   "RHYOLITE, ANDESITE"   80 12

AL SARENA PROJECT
       31S   02E   29   CENTER
       42-51-07N   122-36-35W   93 04

ALCO CREEK
       31S   01E   31   "SE,SE"
       42-44-14N   122-44-14W   93 02

ALICE
       38S   02W   32
       42-12-59N   122-57-59W

ALICE GROUP
       37S   03W   11   NE
       42-22-21N   123-00-58W
       AU   ARGILLITE   80 12

ANDERSON MAGNETITE DEPOSIT
       41S   02W   17
       42-01-16N   122-58-02W

ANDES & HOWARD
       38S   04W   30
       42-14-19N   123-12-48W   AU

ANTELOPE GRAVEL PIT
       36S   01E   30
       42-24-34N   122-44-37W
       SAND & GRAVEL

ANTIMONY LODE
       40S   04W   27   NE
       42-03-58N   123-09-13W
       METAVOLCANIC   80 12

ANTIMONY PROSPECT
       40S   04W   35   "NE,NE,NE"
       42-03-12N   123-07-50W

META-ANDESITE   80 12
APPLEGATE DAM
    40S   03W   08
    42-06-24N   123-05-05W
APPLEGATE LAKE PROJECT # 3
    38S   04W   16
    42-15-33N   123-10-16W
    SAND & GRAVEL
APPLEGATE PROSPECT
    39S   01W   29   "S,S,S"
    42-08-31N   122-51-05W
    CR   GRANULAR METADUNITE   80 12

APPLEGATE RIVER GROUP
    38S   04W   24
    42-15-02N   123-07-47W   STONE(LIMESTONE)
    METASEDIMENTS
ARGONAUT
    35S   04W   02   "NE,NE,NE"
    42-33-52N   123-07-58W
    SCHISTS AND SLATES 80 12
ASH PROSPECT
    34S   01W   01   NW
    42-38-56N   122-46-23W
    "RHYOLITE, RHYOLITIC TUFF BRECCIA"   80 10
ASHLAND MINE
    39S   01E   07   "NW,NW,NW"
    42-11-31N   122-45-24W
    AU AG CU   "QUARTZITE, QUARTZ MICA SCHIST"
        80 12
ASHLAND COAL
    39S   01E   12
    42-11-32N   122-38-21W

ASHLAND GRANITE QUARRY
    40S   01E   01   "SW,NE"
    42-07-16N   122-38-50W
    STONE(GRANITE)   GRANODIORITE   80 12
ASHLAND SHALE DEPOSIT
    39S   01E   12
    42-11-26N   122-39-07W
    CLAY(SHALE)
BAILEY GOLD POCKET
    37S   02W   06   SW
    42-22-32N   122-59-27W
    AU   METASEDIMENTS   80 12

BAILEY PROPERTY
      37S   03W   01   SE
      42-22-30N   122-59-47W
      METASEDIMENTS   80 12
BARR RANCH
      38S   02W   26   "S,NW"
      42-14-17N   122-54-33W
      "METAVOLCANICS, METASEDIMENTS"
BARRON MINE
      39S   02E   23   N
      42-09-58N   122-33-06W
      AU AG ANDESITE BRECCIA   80 12
BASALT QUARRY
      33S   01W   34   "NW,SW"
      42-39-20N   122-48-52W
      STONE(BASALT)
BASALT QUARRY
      33S   02E   09   "NW,SE"
      42-42-48N   122-34-43W
      STONE(BASALT)
BASALT QUARRY
      35S   03E   18   "SE,SW"
      42-31-19N   122-30-54W
      STONE(BASALT)   93 04
BASALT QUARRY
      38S   03W   01   "SE,NW"
      42-17-45N   123-00-23W
      STONE(BASALT)

BASALT QUARRY
      39S   01E   05   SE
      42-12-33N   122-36-05W
      STONE(BASALT)
BASALT QUARRY
      39S   02E   06
      42-12-51N   122-37-20W
      STONE(BASALT)
BASEY CHROMITE DEPOSIT
      40S   01W   32
      42-02-36N   122-50-08W
      CR   DUNITE   80 12
BATZER QUARRY
      36S   01E   05
      42-27-46N   122-43-15W
      STONE
BAXTER LIMESTONE
      37S   03W   02
      42-22-39N   123-00-43W
      STONE

BEAR CANYON
    40S   02E   33   SE
    42-03-00N   122-35-52W
    STONE
BEEMAN LIMESTONE
    37S   03W   11   NE
    42-22-17N   123-00-57W
    METASEDIMENTS   80 12
BEESON MINE
    38S   01E   16   "NE,SW"
    42-15-51N   122-42-33W

BELL AND MANKINS MINE
    38S   02W   20   SW
    42-14-51N   122-58-06W
    AU AG METAVOLCANICS?   80 12
BERTHA CLAIM
    37S   03W   07
    42-21-46N   123-06-23W   AU   "QUARTZITE,
LIMESTONE, ANDESITE"   80 12

BERYLLIUM CLAIM
    34S   01E   10
    42-37-52N   122-40-00W
BERYLLIUM PROSPECT
    39S   02W   34   SE
    42-07-47N   122-55-04W
BETTY
    34S   04W   25
    42-35-17N   123-07-15W   AU
BIG BUCK PROSPECT
    37S   04W   01   CENTER
    42-22-47N   123-07-12W
    AU   GREENSTONE 80 12
BIG CHIEF PROPERTY
    36S   04W   19   S
    42-25-17N   123-12-48W
    AU   "ARGILLITE, DACITE PORPHYRY"
BIG JIM PROSPECT
    34S   02W   19
    42-36-17N   122-58-59W

BIG SHOT
    39S   02W   22
    42-09-46N   122-55-26W
BILL NYE MINE
    36S   03W   33   SE
    42-23-33N   123-03-06W
    AU AG IMPURE QUARTZITES 80 12

BIRDSEYE
      36S   04W   13   CENTER
      42-26-22N   123-07-16W   8C 12
BIRDSEYE CREEK
      37S   04W   11   "NE,NE,NE"
      42-22-31N   123-07-48W
BLACK BOY
      41S   01W   12
      42-00-51N   122-46-10W
      "METAPERIDOTITE, SERPENTINE"   8C 12
BLACK DIAMOND
      39S   01W   15   "SE,NE"
      42-10-43N   122-47-46W

BLACK WARRIOR CLAIM
      40S   01W   32
      42-02-33N   122-50-06W
      CR   PERIDOTITE
BLACK'S PLACER
      38S   03W   05
      42-17-51N   123-04-50W
      AU   CHANNEL GRAVELS   81 05

BLANCHE CLAIM
      36S   03W   24   SW
      42-25-15N   123-00-35W
      TONALITE   80 12
BLOSSOM
      35S   03W   19   NE
      42-31-12N   123-05-50W
      AU AG "ANDESITIC GREENSTONE, SCHIST"
BLOW SNAKE
      41S   03W   01
      42-01-57N   123-00-16W   AU
BLUE BELL
38S   03W   13
      42-16-02N   123-00-15W   AU
BLUE BUCKET
      38S   02W   10   "SW,NE"
      42-16-54N   122-55-16W
      METAVOLCANICS   80 12
BLUE EAGLE
      35S   03W   04
      42-33-00N   123-03-43W
BLUE HILL GROUP
      34S   03W   23
      42-35-59N   123-01-28W

BLUE JAY ANTIMONY
    38S   04W   14
    42-16-00N   123-09-00W   SB
BLUE JAY PROSPECT
    40S   04W   14   "NW,NW"
    42-05-35N   123-08-51W
    SB ", SERPENTINIZED ULTRABASICS" 80 12
BLUE STAR PROSPECT
    37S   04W   14   SE
    42-20-01N   123-08-16W
    METAVOLCANICS   77 05
BOB SWINDLER
    37S   02W   20
    42-20-13N   122-58-05W   STONE
BOITANO ROCK
    37S   02W   32
    42-18-11N   122-58-10W   STONE
BOLLING & KOSTER
    37S   04W   27
    42-19-50N   123-09-28W   AU
BONANZA PROSPECT
    36S   04W   22   CENTER
    42-25-31N   123-09-18W
    METAVOLCANICS   81 05
BONITA
    33S   03W   13
    42-41-59N   123-00-15W
    HG   AMPHIBOLITE 80 10

BONNIE ASBESTOS
    41S   03W   16
    42-00-27N   123-03-37W
    "SCHIST, SERPENTINE?"   80 12
BOOTH
    36S   03W   03
    42-28-13N   123-02-33W
BOWDEN
    36S   02W   19   SW
    42-25-38N   122-59-16W
    AU   TONALITE   80 12
BRADEN MINE
    36S   03W   28
    42-24-25N   123-03-07W
    AU AG "CALCAREOUS HORNBLENDE SCHISTS,
    AMPHIBOLITE, GARNET" 80 12
BRANTER
    38S   04W   25
    42-14-21N   123-07-34W   AU

BRATCHER
　　　39S　　01E　　18　　NW
　　　42-10-42N　　122-45-09W
　　　W　　"HORNFELSED MARBLE, ARGILLITE"
BRATCHER NO. 10
　　　39S　　01W　　25　　NE
　　　42-09-08N　　122-45-48W
　　　QUARTZ DIORITE　　80 12
BRATCHER OPALITE LEDGE
　　　38S　　01E　　24　　NW
　　　42-15-20N　　122-39-16W
　　　"TUFF, ANDESITE"　　80 12
BRICK PILE CHROME
　　　40S　　01W　　06　　NW
　　　42-07-19N　　122-51-54W
　　　METAPERIDOTITE　　80 12
BRICK PILE PROSPECT
　　　40S　　01W　　04　　"S,S,S"
　　　42-06-44N　　122-49-13W
　　　HG　　"AMPHIBOLITES, SILICEOUS
METASEDIMENTS"　　81 01
BRINER PROPERTY
　　　40S　　01W　　25　　SW
　　　42-03-33N　　122-46-26W
　　　QUARTZ DIORITE　　81 05
BRINER QUARRY
　　　38S　　01W　　28　　NW
　　　42-14-33N　　122-49-38W
　　　STONE(LIMESTONE)　　METASEDIMENTS
BRISTOL GRANITE
　　　36S　　03W　　23　　"NE,SE"
　　　42-25-30N　　123-00-53W
　　　STONE(GRANITE)　　GRANODIORITE　　80 12
BRISTOL LIMESTONE
　　　37S　　03W　　06　　NW
　　　42-23-17N　　123-06-34W　　STONE(LIMESTONE)
METASEDIMENTS
BRISTOL SILICA CO. MINE
　　　36S　　03W　　30　　SE
　　　42-24-21N　　123-06-07W
　　　SILICA(LODE)　"GREENSTONE, METASEDIMENTS"
　　　　80 12
BROKEN HEART
　　　38S　　04W　　13
　　　42-16-20N　　123-07-42W　　AU
BROWN BEAUTY
　　　34S　　03W　　14
　　　42-36-50N　　123-01-26W

263

BROWNSBORO #617
      36S   01E   05
      42-28-20N   122-43-25W   STONE
BROWNSBORO CLAY
      36S   01E   11
      42-27-38N   122-39-56W   FRAGMENTAL
TUFF   80 12
BROWNSBORO PIT & PLANT
      37S   02W   25
      42-19-20N   122-53-09W   STONE
BUCK POINT
      39S   02E   26
      42-08-55N   122-33-01W
      ANDESITE   80 12
BUCK ROCK PIT
      36S   03E   10
      42-27-13N   122-27-11W   STONE
BUCKHORN PIT
      40S   02E   02
      42-07-30N   122-32-47W   STONE
BUCKSKIN
      36S   02W   07   "S,S"
      42-26-59N   122-59-11W
      AU AG METAVOLCANIC   80 12
BULL OF THE WOODS PROSPECT
      36S   03W   15   NE
      42-26-44N   123-02-02W
      AU AG DIORITE (GABBRO)   80 12
BULL RUN
      38S   02W   33
      42-13-29N   122-56-36W
BULLARD
      39S   04E   34   S
      42-08-09N   122-19-51W   STONE
BULLFROG PLACER
      36S   04W   22
      42-25-20N   123-09-39W   AU
BUNCE PROSPECT
      33S   04W   09   CENTER
      42-43-44N   123-11-06W
      "METAGABBRO, SERPENTINITE"   80 12

BURDIC GROUP
      39S   01W   13   N
      42-11-06N   122-46-28W   HORNFELSED
      METASEDIMENTS   80 12

BURNT CREEK PIT
      38S   02E   27
      42-13-49N   122-33-43W   CLAY
BURNT CREEK QUARRY
      38S   03E   30   SE
      42-14-11N   122-30-14W   CLAY
BURNT PINE 1-2-3
      36S   04W   28
      42-24-46N   123-10-47W   AU

BURRIS PROSPECT
      39S   01W   11   E
      42-11-33N   122-47-08W   QUARTZITE
BUSH RANCH
      37S   02E   09   NE
      42-22-05N   122-34-58W   RED TUFF
BUTTE FALLS
      33S   01E   27
      42-40-23N   122-41-01W
CALL OF THE WILD
      35S   02E   29   "SE,SE"
      42-29-37N   122-36-12W
      FLOW BRECCIA   81 01
CALUMET
      37S   04W   05   "SE,SW"
      42-22-39N   123-11-55W
      AU AG   80 12
CANYON CREEK MINING CO.
      33S   01W   31
      42-39-30N   122-51-50W
      RHYOLITO FLOWS AND TUFFS
CAPITOL HILL PROSPECT
      36S   04W   09   SE
      42-27-04N   123-10-47W
      METAVOLCANICS   81 01
CARBERRY CREEK GROUP
      40S   04W   19
      42-04-26N   123-13-03W
CARBONATE
      35S   03W   17   W
      42-31-39N   123-05-12W
      AU AG DIORITE ? ?   81 01
CARLTON
      33S   01E   23   "SE,NE"
      42-41-23N   122-39-34W
      PUMICE   NUE ARDENTE DEPOSIT   81 01

CARTINELL PROSPECT
36S   04W   09   CENTER
42-27-10N   123-10-46W
METAVOLCANICS (ANDESITE)81 01
CASCADE VIEW PROSPECT
35S   03W   35   SE
42-29-16N   123-02-00W   AU AG "SLATE,
METAVOLCANICS"   81 01
CASEY
38S   01W   32   SW
42-12-56N   122-51-08W
"QUARTZITES, SILICIC METAVOLCANICS,
METARHYOLITE"   81 01

CASS RANCH
39S   01W   29   "S,S,S"
42-08-28N   122-51-10W
CR   DUNITE METAPERIDOTITE
METAPERIDOTITE   80 12
CEDAR
33S   04W   21
42-41-16N   123-10-47W
CHANCE DISCOVERY CLAIM
41S   01W   10
42-01-27N   122-48-16W
CR   DUNITE METAPERIDOTITE
METAPERIDOTITE   80 12
CHARLOTTE
40S   04W   05   NW
42-07-25N   123-12-14W
SHEARED METAVOLCANICS   81 01
CHISHOLM
34S   02W   17   "S,S,S"
42-36-31N   122-57-51W
HG   "SANDSTONE, AMPHIBOLITE" 80 10
CHROME KING
34S   04W   03
42-39-02N   123-09-30W   CR
CHROME PROSPECT
34S   02W   07   "E,E,E"
42-37-48N   122-58-26W
ULTRAMAFICS 81 05
CHROMITE OCCURRENCE
39S   01W   32
42-08-22N   122-50-56W

CHROMITE OCCURRENCE
      34S   04W   20
      42-35-60N   123-11-55W

CINNABAR MOUNTAIN
      34S   02W   34
      42-34-26N   122-55-50W
      HG   SANDSTONE   80 10
CLAY PIT
      37S   01E   15   SW
      42-21-03N   122-41-29W   CLAY(SHALE)
CLIMAX GOLD AREA
      38S   01E   06
      42-17-43N   122-38-01W
      TUFF BRECCIA
CLOUDY DAY QUARTZ CLAIM
      40S   04W   16   W
      42-05-43N   123-11-19W
      AU   GREENSTONE 81 01
COAL PROSPECT
      39S   02E   07   NE
      42-11-37N   122-38-20W
      "SHALE, SANDSTONE" 80 12
COLLINS
      40S   02W   35
      42-03-02N   122-54-15W
COLPITTS-TUER-PEMINCER
      36S   02W   28
      42-24-55N   122-56-29W
      SAND & GRAVEL
COLUMBINE CLAIM
      39S   01W   14
      42-10-39N   122-46-60W
      DIORITE   81 01
CONGER QUARTZ
      36S   02W   18   SW
      42-26-09N   122-59-27W
      GRANODIORITE?   81 01
COOK MINE
      37S   04W   13
      42-21-00N   123-07-12W   AU
COON CREEK CLAIMS
      37S   02E   20   "SE,SE"
      42-20-39N   122-36-04W   RED TUFF   80 12

COPELAND CREEK QUARRY
      30S   04E   07
      42-58-46N   122-23-11W   AU

COPPER OCCURRENCE
        34S   04W   02
        42-39-03N   123-08-45W   AMPHIBOLITE
COPPER OCCURRENCE
        40S   02W   11   "S,S,S"
        42-05-54N   122-53-56W   DIORITE?
COPUS
        40S   03W   02   CENTER
        42-07-12N   123-01-18W
CORPORAL G
        35S   03W   19   S
        42-30-26N   123-05-54W
        AU AG SLATY QUARTZITES   81 01

COSTER & CATTON'S CLAIM
        37S   04W   21   SW
        42-20-08N   123-11-02W   GREENSTONE
        81 01
CRACKER JACK CLAIM
        39S   01W   14   "NE,SW"
        42-10-34N   122-47-13W
        METAMORPHOSED SHALE   81 01
CRATER COAL CO
.   37S   01W   36   W
        42-18-36N   122-46-23W
        COAL   "SHALE, SAND STONE, CONGLOMERATE
        STONE, CONGLOMERA"
CROUCH
        33S   04W   16
        42-41-49N   123-11-18W
        CR   SERPENTINE
CUTTER CLAIM
        41S   04W   09
        42-00-57N   123-10-48W
CYNTHIA TREMOLITE
        33S   04W   33
        42-39-41N   123-10-44W
        SERPENTINE
DAVE FORCE MINE
        34S   02W   20
        42-36-23N   122-57-20W
        HG   SANDSTONE   80 10
DAVIES PLACER
        38S   03W   04
        42-17-41N   123-04-02W
        AU SAND & GRAVEL

DAVIS LEDGE PROSPECT
    37S   03W   13
    42-21-21N   123-00-23W
    AU   81 01
DAWN MARIE CLAIM
    33S   01W   19
    42-41-18N   122-51-57W   TUFF   80 12
DAWSON MERCURY PROSPECT
    34S   01E   02
    42-38-34N   122-40-14W
    RHYOLITE   81 01

DAWSON-MURPHY PIT
    40S   04W   20   SW
    42-04-26N   123-12-23W
    SAND & GRAVEL
DEAD INDIAN CLAY
    38S   03E   19
    42-15-12N   122-30-43W
    CLAY   ALTERED RHYOLITE   81 01
DEAD INDIAN ROCK QUARRY
    38S   02E   33
    42-13-16N   122-35-22W
    STONE
DEMOSS
    37S   02W   17   SE
    42-21-00N   122-57-27W
    STONE(GRANITE)
DEYOUNG PIT
    38S   01E   31
    42-13-41N   122-45-05W
    SAND & GRAVEL

DIXIE QUEEN
    37S   03W   18   NW
    42-21-19N   123-06-12W
    CALCEROUS ARGILLITE   81 01
DONALDSON
    33S   01W   35   SE
    42-39-05N   122-46-47W
    RHYOLITE FLOWS AND CRYSTAL-LITHIC TUFFS
        80 10
DORIS LEDGE
    37S   03W   13   SW
    42-20-54N   123-00-15W
    GRANODIORITE   81 01

DOUBLE H
      36S   04W   27   "NW,SW"
      42-24-28N   123-09-55W
      AU   METASEDIMENTARY 81 01
DOUBLE JACK
      39S   01W   14   NW
      42-10-564   122-47-276
      AU AG "HORNFELSED MAFIC VOLCANICS,
      METASEDIMENTS" 80 12
DRAGON TAIL
      33S   04W   08
      42-42-37N   123-12-04W
DRAKE-YOUTSEN PROJECT 1
      38S   04W   07
      42-16-47N   123-13-06W   SAND & GRAVEL
DRAPER HOMESTAKE
      37S   03W   30   CENTER
      42-19-25N   123-06-03W   81 01

DUCHARME NO. 6
      36S   04W   13   NW
      42-25-59N   123-07-22W
      ARGILLITE   81 02
DUNROMIN
      36S   03W   36   SW
      42-23-33N   123-00-31W
      AU AG QUARTZ DIORITE   81 01
EAGLE
      36S   03W   25
      42-24-19N   122-59-40W
      AU AG
      ARGILLITE META-ANDESITE META-ANDESITE
EDWARDS & GARRISON
      41S   02W   17   NE
      42-00-34N   122-57-34W
      AU   CHLORITE SCHIST   81 01
EL SONORA
      38S   04W   34   N
      42-13-28N   123-09-35W
      AU   ARGILLITE   81 01
ELK CREEK QUARRY
      33S   01E   18   "SE,SE"
      42-41-41N   122-44-03W
      STONE(ANDESITE)   93 02
ELKHORN CLAIMS
      32S   02W   21
      42-46-29N   122-56-38W
      MICA SCHISTS AND AMPHIBOLITES OF APPLEGATE
      GROUP 81 01

ELLINGER ASBESTOS
    41S   02W   13
    42-00-20N   122-53-16W
    ASBESTOS   METASERPENTINITE
ENSELE QUARRY
    38S   02W   06   NW
    42-17-50N   122-59-19W   STONE(LIMESTONE)
    81 01
ERICKSON PROSPECT
    35S   03W   06   E
    42-33-09N   123-05-53W
    MICA-SCHIST QUARTZITE QUARTZITE
EVANS CREEK PROSPECT
    34S   02W   17
    42-36-35N   122-57-35W
    QUARTZITE MICA SCHISTS MICA SCHISTS
EVERGREEN
    37S   03W   18
    42-21-07   123-06-02W   AU
FABRICATORS QUARRY
    33S   01E   32   "NE,NW"
    42-39-43N   122-43-46W   STONE
FAKIR
    39S   04W   04
    42-12-28N   123-10-46W   AU
FEDERAL
    39S   03W   13
    42-10-24N   123-00-27W   AU
FIRST HOPE
    37S   04W   07   SW
    42-21-47N   123-13-29W   AU
    "GREENSTONE, PORPHYRITIC ANDESITE"
FLEMING
    38S   03W   23   S
    42-14-33N   123-01-27W
    AU AG SLATE 80 12
FLOAT OCCURRENCE
    36S   03W   13   SW
    42-26-09N   123-00-28W
FLUME GULCH PROSPECT
    36S   04W   01   NE
    42-28-25N   123-06-47W
    QUARTZITE 81 01
FLYING PORCUPINE QUARRY
    41S   02E   03   "NW,NE"
    42-02-19N   122-34-19W   STONE(BASALT)

FOOTHILLS BLVD. PIT
      36S   04W   18
      42-26-20N   123-13-16W   STONE

FOOTS CREEK PLACERS
      37S   03W   32
      42-18-56N   123-04-37W
      AU AU COLLUVIUM AND CHANNEL GRAVELS
         81 05

FOOTS CREEK TUNGSTEN
      37S   04W   13   "SE,NW"
      42-21-18N   123-07-37W   HORNFELSED
METASEDIMENTS

FOREST CREEK PLACER
      38S   03W   14
      42-15-54N   123-01-53W
      AU   STREM AND CHANNEL GRAVELS

FORSET CREEK
      38S   03W   04
      42-17-27N   123-03-11W   AU

FORT KNOX
      33S   04W   28
      42-40-37N   123-10-26W   CU   GREENSTONE
      (BASALT)   81 01

FORTY - NINE DIGGINGS
      38S   01E   31
      42-13-20N   122-44-19W
      AU HG QUAT ALLUVIUM AND CRET
      CONGLOMERATE AND SANDSTONE   81 01

FOSTER CREEK SITE
      30S   04E   06   NE
      42-59-36N   122-23-18W   AU

FOUR CORNERS
      41S   01E   08
      42-01-27N   122-43-45W

FOUR SPOT
      38S   04W   19
      42-15-16N   123-13-10W   AU

FOX PROSPECT
      37S   02E   17
      42-20-58N   122-36-19W
      ANDESITE AND TUFF 80 12

FRANK DAILEY GRAVEL PIT
      37S   04W   15
      42-21-28N   123-09-04W   SAND & GRAVEL

FREE GOLD
      39S   02W   14
      42-10-48N   122-54-10W

FREEMAN FELDSPAR
    39S   01E   27   SW
    42-08-45N   122-41-25W
    PEGMATITE   81 01

FREEWAY TUNGSTEN
    40S   02E   08   "CENTER,N"
    42-06-30N   122-36-42W
    HORNFELSED SEDIMENTS MARBLE   81 01

FROG
    40S   03W   31
    42-03-01N   123-06-22W
FROG CREEK QUARRY
    38S   02E   33
    42-13-29N   122-35-34W
    STONE
FROST
    34S   02W   20
    42-36-02N   122-57-52W
GAINES PROPERTY
    33S   01W   20   SW
    42-40-504   122-51-082
GALLS CREEK GROUP
    36S   03W   21   CENTER
    42-25-35N   123-03-43W
    LIMESTONE   81 02
GALLS CREEK PLACERS
    37S   03W   04   NE
    42-23-06N   123-03-21W
    AU   COLLUVIUM AND CHANNEL GRAVELS
        81 05
GAMMILL PROSPECTS
    37S   04W   18
    42-21-13N   123-12-56W
    "ARGILLITE, QUARTZITE"
GC&E PLACER CLAIM
    36S   03W   12
    42-27-14N   123-00-15W
GEMSTONE MINE
    41S   04E   08   "W,SW"
    42-00-50N   122-23-05W   GEMSTONE(AGATE)
GEORGE A MCLEAN
    36S   01W   02
    42-28-06N   122-47-06W
GLADE VIEW
    39S   01W   29
    42-08-43N   122-51-00W

GLEN DITCH PLACER
        37S   04W   27
        42-19-02N   123-09-39W   AU
GLORY HOLE
        33S   04W   17
        42-42-20N   123-11-43W
        CR   SERPENTINE   81 02

GOLD CHLORIDE
        35S   04W   25   NE
        42-30-13N   123-06-45W
        AU AG QUARTZITES QUARTZ MICA SCHISTS
        QUARTZ MICA SCHISTS
GOLD COIN PROSPECT
        38S   03W   02   NW
        42-18-06N   123-01-56W   81 01
GOLD CYCLE
        40S   02E   17   NE
        42-05-36N   122-36-27W
        QUARTZ DIORITE   81 01
GOLD DUST PLACER
        37S   03W   31   "NW,NE"
        42-18-53N   123-06-10W
        METAVOLCANICS   81 05
GOLD HILL MINE
        36S   03W   14
        42-26-36N   123-01-06W
        AU PT PYROXENITE 81 05
GOLD HILL PLACER
        36S   03W   17
        42-26-18N   123-04-35W
        AU   STREAM GRAVELS   81 05
GOLD MINE
        33S   04W   10
        42-42-42N   123-09-16W   AU
GOLD MINE
        33S   04W   21   "SE,SE"
        42-40-57N   123-10-24W   AU
GOLD MINE
        37S   03W   36   NE
        42-18-27N   122-59-57W   AU   93 04
GOLD MINE
        38S   03W   10
        42-16-30N   123-02-21W   AU
GOLD MINE
        38S   03W   15   NW
        42-16-06N   122-55-46W   AU   93 04

GOLD MINE
    39S   03W   03   SE
    42-12-03N   123-02-25W   AU
GOLD MINE
    41S   02W   10
    42-01-27N   122-55-21W   AU
GOLD NOTE
    33S   04W   30   "W,W,W"
    42-40-21N   123-13-38W   AU CU AG
    "SLATE, METAVOLCANICS"   81 01
GOLD PAN
    40S   04W   02   W
    42-07-19N   123-08-53W   AU AG
    "GREENSTONE, ARGILLITE"   80 12
GOLD PLATE
    35S   03W   17
    42-31-31N   123-05-18W   AU AG SCHIST
MARBLE   81 01
GOLD PROSPECT
    34S   04W   23
    42-36-07N   123-08-53W
GOLD PROSPECT
    36S   03W   32   NE
    42-24-01N   123-04-36W
    METAVOLCANICS   81 05
GOLD PROSPECT
    37S   04W   01   "W,W,W"
    42-22-55N   123-07-50W
    METAVOLCANICS   81 05
GOLD PROSPECT
    37S   03W   15   "NE,SW"
    42-21-04N   123-02-43W
    METAVOLCANICS   81 05
GOLD PROSPECT
    37S   03W   22
    42-20-35N   123-02-28W
    METAVOLCANICS   81 05
GOLD PROSPECT
    38S   04W   14   "E,NW"
    42-16-11N   123-08-33W
    METAVOLCANICS   81 05
GOLD PROSPECT
    38S   04W   15   E
    42-15-59N   123-09-21W
    METASEDIMENTS   81 05
GOLD PROSPECT
    38S   04W   24   "SE,SE,SE,NE"
    42-14-59N   123-06-44W   METAVOLCANICS
    81 05

GOLD PROSPECT
>   38S   03W   02   "SW,NW"
>   42-17-49N   123-01-50W   METAVOLCANICS
>   81 05

GOLD PROSPECT
>   39S   04W   25   S
>   42-08-43N   123-07-44W

GOLD PROSPECT
>   40S   02E   14   NW
>   42-05-41N   122-33-19W   VOLCANICS   81 05

GOLD RAY GRANITE
>   36S   02W   18   SW
>   42-26-19N   122-59-12W
>   STONE(GRANITE)   CLAY   CLAY   81 01

GOLD RIDGE
>   37S   03W   03   NE
>   42-23-15N   123-02-02W   AU AG "QUARTZITE,
>   ANDESITE"   81 01

GOLD STANDARD
>   37S   03W   25
>   42-19-22N   123-00-20W

GOLDEN BUTTE
>   39S   01E   27
>   42-09-11N   122-40-56W

GOLDEN PLUME
>   39S   04W   27   NW
>   42-09-09N   123-09-46W
>   METAVOLCANICS ?

GOLDEN ROAD
>   33S   04W   35
>   42-39-00N   123-08-43W   CU AU   AMPHIBOLITE
>   81 01

GOLDEN STAR
>   40S   04W   03
>   42-06-48N   123-10-21W   AU   ANDESITE
>   PORPHYRY   80 12

GOLDEN THISTLE
>   38S   03W   21
>   42-15-13N   123-03-48W   AU

GOOD FRIDAY
>   38S   01W   36
>   42-13-18N   122-45-58W   81 01

GRAND COVE PROSPECT
>   35S   02E   29
>   42-30-00N   122-36-25W

GRANGE GULCH
     38S   04W   24
     42-15-16N   123-07-15W   AU
GRANITE
     37S   02W   07
     42-21-45N   122-58-53W   STONE
GRANITE
     39S   01E   06
     42-12-46N   122-44-19W   STONE
GRANITE QUARRY
     35S   01W   21
     42-30-55N   122-49-18W   STONE(GRANITE)
GRANITE QUARRY
     37S   02W   19   "SE,NW"
     42-20-22N   122-59-11W   STONE(GRANITE)
GRANITE QUARRY
     37S   02W   19   "SE,SW"
     42-19-52N   122-59-13W
     "STONE(GRANITE), SHALE"   93 04
GRANT POWELL PROSPECT
     35S   03W   32   SW
     42-28-55N   123-05-18W
     METASEDIMENTS   81 01
GREAT I AM
     38S   04W   31   NW
     42-13-36N   123-13-34W   AU AG GREENSTONE
     80 12
GREB
     36S   01W   04   NE
     42-28-27N   122-49-07W   STONE
GRERS SHALE PIT
     36S   01W   03
     42-27-48N   122-47-50W   STONE
GROUSE CREEK QUARRY
     39S   03W   25
     42-09-12N   122-59-42W   STONE
GROWLER
     39S   01W   13   NE
     42-10-55N   122-45-39W
     AU AG GRANODIORITE   81 01
GRUBSTAKE
     41S   02W   09   "S,SW"
     42-00-55N   122-56-53W
     AU   "QUARTZ-MUSCOVITE SCHISTS, GRAPHITIC
     SCHISTS, ACTIN"   80 12

H.P. ANTIMONY
        40S   01E   32   SE
        42-02-34N   122-43-13W   QUARTZ
        DIORITE   81 01
HALE QUARRY
        36S   01W   04
        42-28-15N   122-49-27W   STONE
HALL PROSPECT
        39S   01E   07   "W,SE"
        42-11-11N   122-44-42W   W
        GRANODIORITE/SCHIST   81 01
HAMILTON
        39S   01E   23
        42-09-32N   122-40-15W   STONE(GRANITE)
HAMILTON-TAYLOR RANCH PLACER
        38S   03W   33
        42-13-17N   123-03-50W   AU
HANCOCK CLAIMS
        37S   04W   09
        42-21-59N   123-10-30W   GABBRO   81 01
HANSEN COAL MINE
        37S   01W   03
        42-22-29N   122-48-56W
        SHALE 81 01

HANSEN SITE
        36S   01W   03
        42-28-21N   122-48-48W   STONE
HANSON RANCH
        34S   02W   02   CENTER
        42-38-31N   122-54-19W
        RHYOLITE FLOWS AND TUFFS 81 05
HARMONY GROUP
        38S   01E   18
        42-16-03N   122-44-55W
HARTH AND RYAN
        36S   04W   33
        42-23-45N   123-10-44W
        GREENSTONE (ANALYSIS GIVEN)   81 01
HASKINS & TRAVERSO
        41S   02W   06   NW
        42-02-27N   122-59-02W
        META-ANDESITE   81 01
HATCH QUARRY
        36S   04W   10
        42-27-10N   123-10-02W
        STONE

HAZEL GROUP
        36S   04W   27   SW
        42-24-17N   123-09-55W
        AU   "LIMESTONE, SLATE" 80 12
HEATH
        36S   03W   12   SW
        42-26-53N   123-00-31W
        HORNFELSED METAVOLCANICS   81 01
HEMATITE
        41S   02W   06
        42-02-27N   122-59-03W
        AU   "ARGILLITE, META-ANDESITE"   80 12

HENRY AND BETTY SHANKNIS DIGGINGS
        35S   03W   25
        42-30-00N   123-07-25W
        UMPQUA FORMATION
HIDDEN TREASURE
        36S   04W   16   NW
        42-26-26N   123-11-05W
        AU   "ARGILLITE, IMPURE QUARTZITE,
        METAVOLCANICS"   80 12

HIGH BANKS PIT
        37S   02W   10
        42-22-35N   122-54-49W
        SAND & GRAVEL
HIGHLAND CLAIM
        37S   04W   22   SW
        42-20-02N   123-09-52W
        SCHIST 81 01
HOLLOW TREE
        38S   03W   36
        42-13-12N   123-00-16W   AU
HOMESTAKE CLAIM
        36S   02E   05
        42-28-02N   122-37-01W
        REDDISH TUFF 80 12
HOPE MINING CLAIM
        40S   03W   31   "SE,SE"
        42-02-50N   123-08-29W
        AU   SCHIST 81 01
HOPELESS
        38S   02E   09   "N,SW"
        42-16-42N   122-35-44W
        TUFF   80 02
HORSESHOE
        38S   03W   29
        42-14-00N   123-04-51W

HORSESHOE CLAIMS
    40S   01W   05
    42-07-32N   122-51-02W
    CR   METASERPENTINITE 80 12
HUMBUG CREEK PLACERS
    38S   04W   11
    42-16-09N   123-08-20W
    AU   CHANNEL GRAVELS 81 05
HUMDINGER MINE   39S
    01W   23   NW
    42-10-08N   122-47-26W
    METASEDIMENTS   81 01

HYATT ROCK QUARRY
    39S   03E   20
    42-09-24N   122-29-16W   STONE
IDEAL CEMENT CO.
    36S   03W   12
    42-27-24N   123-00-30W   STONE(LIMESTONE)
CLAY CEMENT
IDEAL CEMENT QUARRY
    36S   03W   16   SW
    42-26-21N   123-03-59W   SHALE
INDIAN CREEK QUARRY
    34S   01W   22   "W,SE"
    42-35-47N   122-48-14W   STONE(BASALT)
IRON CROWN
    38S   02W   33
    42-13-10N   122-56-50W   AU
IRON HAT
    41S   01W   30   SE
    42-03-28N   122-51-33W   AMPHIBOLITE

IRON MOUNTAIN PLACER
    35S   03W   24
    42-30-42N   122-59-54W   AU
IRON OCCURRENCE
    40S   01W   28   "NE,NW"
    42-04-10N   122-49-30W   FERRUGINOUS
CHERTS
IRWIN MOLYBDENUM PROSPECT
    36S   04W   16   NE
    42-26-35N   123-10-28W
    BI   89 08
JACKSON COUNTY
    33S   02W   27
    42-44-07N   122-54-30W   STONE

JACKSON EXCHEQUER PLACER
        40S   04W   19   "SE,SE"
        42-04-22N   123-12-50W
        AU   TERRACE GRAVEL   81 05
JACKSONVILLE
        37S   02W   19
        42-20-04N   122-59-04W   STONE(GRANITE)

JACKSONVILLE BRICK & TILE
        37S   02W   31   "SW,NW"
        42-18-35N   122-59-23W   CLAY   81 01
JACKSONVILLE MINING & MILLING
        37S   03W   25
        42-19-06N   123-00-38W
JACKSONVILLE OLD RES
        37S   03W   25
        42-19-38N   123-00-34W   SAND & GRAVEL
JACKSONVILLE PLACER 5
        37S   02W   32
        42-19-02N   122-58-11W
        AU   CHANNEL GRAVELS   81 05
JAY BIRD
        40S   04W   14
        42-05-30N   123-08-30W
        SB   METAMORPHIC & VOLCANIC ROCKS
JAY'S PIT
        38S   01W   18
        42-15-40N   122-51-21W
        STONE
JELDNESS AND RHODES
        41S   02W   06   NE
        42-02-09N   122-58-45W
        ALTERED ULTRAMAFIC ROCKS NEAR CONTACT
        WITH SCHIST   80 02
JENCO/LITTLE BUTTE
        36S   01W   04
        42-28-04N   122-49-10W   STONE
JIM DANDY
        37S   03W   17
        42-21-16N   123-04-49W   AU
JOHN HENDERSON
        39S   02W   04
        42-12-05N   122-56-59W   SAND & GRAVEL
JOHN PEAK SITE
        37S   02W   17   "NW,NW"
        42-21-15N   122-58-20W   STONE(GRANITE)

JOHNSON
        36S   04W   23   SE
        42-25-18N   123-08-01W
        LIMESTONE   81 01
JOLLY PROSPECT
        37S   03E   08
        42-22-04N   123-04-51W
        BASALT   81 02
JONATHAN WAY
        37S   02W   19
        42-20-07N   122-58-34W   STONE(GRANITE)
JUBY LODE
        40S   03W   31   SE
        42-02-33N   123-06-29W   AMPHIBOLITE
        81 02
JUDSON CLAIM
        36S   04W   25   "SE,NE"
        42-24-46N   123-07-00W   GREENSTONE
& LIMESTONE 81 01
JUNCTION CLAIMS
        41S   01E   08
        42-01-34N   122-44-25W   GRANODIORITE
JUNEBURG PROSPECT
        38S   04W   06   "SE,SE"
        42-17-24N   123-12-38W   DIORITE
KANE CREEK PLACERS
        36S   03W   35   N
        42-23-56N   123-01-17W
        AU AG STREAM GRAVEL   81 05
KENDALL BAR
        36S   02W   16   SW
        42-26-12N   122-57-03W   SAND & GRAVEL
KINDSCHI QUARRY
        33S   01E   34   "NE,NW"
        42-39-42N   122-41-25W   STONE(ANDESITE)
KING MOUNTAIN
        33S   04W   18
        42-42-22N   123-13-33W
        CR   SERPENTINE PERIDOTITE   81 01
KUBLI
        37S   03W   05   NW
        42-23-15N   123-05-04W
        AU AG METAVOLCANICS AND METASEDIMENTS
               80 12

LADY SLIPPER PROSPECT
>37S   03W   07
>42-21-57N   123-05-50W
>"METAVOLCANICS, TUFFS AND
>AMYGDALOIDAL LAVAS"   80 12

LADY SUE
>34S   04W   07   "E,SW"
>42-37-32N   123-13-12W   SHALE 81 01

LANCE PLACER
>37S   04W   22   SE
>42-20-20N   123-09-11W
>AU   CHANNEL GRAVEL   81 05

LANHAM/LYMAN BAR
>36S   03W   11
>42-27-24N   123-00-59W   SAND & GRAVEL

LAST CHANCE GROUP PLACER
>33S   04W   15
>42-42-31N   123-09-33W   AU

LAST CHANCE MINE
>36S   03W   33   NE
>42-23-40N   123-03-58W   ANDESITE

LAST CHANCE MINE
>40S   04W   06   NE
>42-07-25N   123-12-42W

LAST CHANCE PROPERTY
>33S   04W   03
>42-44-03N   123-10-02W

LAST RESORT
>39S   02W   04   "E,W"
>42-12-27N   122-56-37W   AU AG

LAUREL CLAIMS
>40S   04W   09   "SW,SW"
>42-05-59N   123-11-19W
>SCHISTOSE METAVOLCANICS

LAWRENCE MINE
>36S   03W   33   "SE,SE"
>42-23-37N   123-03-16W

LAYTON MINE
>38S   04W   20   "SW,SW,SW"
>42-14-56N   123-12-28W
>AU   TERRACE AND CHANNEL GRAVELS

LEE
>35S   03W   06
>42-33-37N   123-06-23W
>MN   QUARTZITE QUARTZ MICA SCHIST QUARTZ
>MICA SCHIST   81 01

LEMASTER GRAVEL PIT
        37S   04W   15
        42-21-06N   123-09-04W   SAND & GRAVEL
LENHERT PLACER
        35S   03W   07
        42-32-24N   123-05-40W   AU
LIBERTY ASBESTOS
        32S   04W   36
        42-44-40N   123-06-58W
        ASBESTOS   METAVOLCANIC   81 01
LICK CREEK COPPER
        35S   01W   35
        42-28-43N   122-39-50W
LIKENS PROSPECT
        36S   04W   34
        42-24-12N   123-10-06W   GREENSTONE
81 01
LIMESTONE QUARRY
        36S   04W   13   "SE,SE"
        42-26-04N   123-06-48W
        LIMESTONE METASEDIMENTS81 01
LIMESTONE QUARRY
        36S   04W   36
        42-24-02N   123-07-33W   STONE(LIMSTONE)

LININGER SITE
        37S   02W   19   "SE,SW"
        42-20-03N   122-59-21W   STONE(GRANITE)
LINN ROAD SHALE PIT
        36S   02W   35
        42-24-04N   122-54-34W   STONE
LITTLE APPLEGATE PLACERS
        39S   02W   20   "SW,SW,SW"
        42-09-22N   122-58-24W   AU   81 05

LITTLE ARCTIC
        33S   04W   08   SW
        42-42-43N   123-12-07W
        AU AG "SERPENTINE, GREENSTONE, GREENSTONE"
LITTLE JOHNNY
        36S   03W   28
        42-24-28N   123-03-52W   GREENSTONE-
GRANITE   81 01
LITTLE MOON
        39S   01W   25
        42-09-01N   122-46-01W
LIVELY LIMESTONE MINE
        37S   03W   11
        42-22-56N   123-01-13W   STONE(LIMESTONE)
284

LLANO DE ORO
      40S   04W   17   CENTER
      42-04-30N   123-11-49W
      SCHIST AND ANDESITE   81 01
LONE EAGLE PROSPECT
      35S   03W   29   SE
      42-29-31N   123-04-28W
      AU AG "QUARTZITE, ANDESITE"   81 01
LONE PINE DENDRITE
      41S   42E   07   SW
      42-00-53N   122-24-04W   GEMSTONES(AGATE)
LONE PINE MINE
      38S   03W   15   "N,N"
      42-16-11N   123-02-11W
      SAND & GRAVEL   "ARGILLITE , SERPENTINE,
      SERPENTINE" 81 01
LONG BRANCH
      34S   02W   24   CENTER
      42-36-02N   122-53-08W
      VOLCANIC ROCKS   81 02
LOOKOVER
      39S   02W   15
      42-10-47N   122-55-19W
LOST
      38S   03W   01
      42-17-43N   123-00-22W   AU
LOST CABIN
      36S   03W   18
      42-26-30N   123-06-02W
LOST CREEK DAM QUARRY
      33S   01E   35   "S,NE"
      42-39-36N   122-39-43W
      STONE(ANSESITE)   93 02
LOST CREEK QUARRY
      32S   03E   29
      42-45-13N   122-29-12W   STONE
LOVE PROSPECT
      40S   01W   33
      42-02-49N   122-49-23W
      METAPERIDOTITE   80 12
LOWRY ANTIMONY MINE
      40S   04W   25   NW
      42-03-53N   123-07-31W
      SB   "META-ANDESITE, ARGILLITE"   81 01
LUCKY 13 PROSPECT
      38S   02E   09
      42-16-57N   122-35-35W
      LITTLE BUTTE VOLCANICS   81 02

LUCKY BART
        35S   03W   29   SW
        42-30-15N   123-05-22W   AU AG "SLATE,
MICACEOUS QUARTZITES"   80 12
LUCKY FRIDAY
        40S   04W   17   SE
        42-05-22N   123-11-39W
        AU   META-ANDESITE   81 01
LUCKY GIRL
        40S   03W   05
        42-07-20N   123-05-00W
LUCKY GULCH
        39S   02W   12
        42-11-46N   122-52-49W
LUCKY KING
        37S   02W   07
        42-22-07N   122-58-59W   AU

LUCKY STRIKE PROSPECT
        37S   04W   14   NE
        42-21-25N   123-08-31W
        METAVOLCANICS LIMESTONE LIMESTONE

LULL GRANITE
        37S   02W   08
        42-22-18N   122-57-34W   STONE(GRANITE)
LULL QUARRY
        36S   02W   08
        42-27-13N   122-57-42W   STONE
LYONS GULCH
        37S   04W   01
        42-22-42N   123-06-53W
MACE BAR
        36S   02W   15
        42-26-18N   122-55-51W   SAND & GRAVEL
MAID OF THE MIST
        39S   04W   04   SE
        42-12-18N   123-10-30W   GREENSTONE
        80 12
MAMMOTH LODE
        32S   02W   28   "W,W,W"
        42-45-47N   122-57-02W   AMPHIBOLITE

MAMMOTH PROSPECT
        38S   02E   09   "NW,NE"
        42-17-08N   122-35-15W
        SILICIC TUFF   81 02

MANGANESE PROSPECT
    35S   04W   30
    42-29-43N   123-06-28W   81 05
MANKINS PROSPECT
    38S   02W   19   NE
    42-15-07N   122-58-44W   81 01
MANSFIELD
    36S   02W   30   CENTER
    42-24-48N   122-58-52W
    AU   METAVOLCANICS   81 01
MAPLE GULCH 34S   03W   27
    42-35-06N   123-30-28W
MARK I GROUP
    40S   01E   27
    42-03-43N   122-41-20W
    PEGMATITE   80 12
MARTIN LIMESTONE
    36S   04W   21   SE
    42-25-25N   123-10-13W
    STONE(LIMESTONE)   81 01
MATTERN
    38S   01E   31   SE
    42-13-06N   122-44-31W
    "DIORITE, QUARTZ DIORITE" 80 12
MAX CLAIM
    39S   02W   21
    42-09-59N   122-57-01W
    META-ANDESITE
MC LEMORE & HAMPSON'S
    37S   03W   07   SE
    42-21-45N   123-05-34W
    AU   GREENSTONE 81 01
MC MAHON'S CLAIM
    37S   03W   06   "NW,SW"
    42-22-48N   123-05-52W
    "GREENSTONE, QUARTZITE" 81 01
MC TIMMONS MINE
    33S   04W   19   "NW,SE"
    42-41-14N   123-12-53W
    AU   METABASALT 81 01
MCCAULEY PROSPECT
    35S   03W   32   "SW,SW"
    42-28-53N   123-01-41W
    METAVOLCANICS
MCCULLEY PROSPECT
    34S   02W   18
    42-36-39N   122-58-45W

MCGREW PROSPECT
    40S   03W   15
    42-05-40N   123-02-45W
MEADOWS
    34S   02W   21   "S,NW"
    42-36-22N   122-57-37W
    SANDSTONE ARGILLITE   90 12
MEDCO QUARRY
    34S   01E   12
    42-37-18N   122-38-32W
    STONE(ANDESITE)   93 02
MEDFORD READY MIX PIT AND CRUSHER
    37S   01W   25   NE
    42-19-37N   122-45-41W
    SAND & GRAVEL CLAY
MEE PLACER
    38S   04W   28
    42-14-34N   123-10-21W   AU
MERCURY OCCURRENCE
    40S   04W   20
    42-04-37N   123-12-09W
MERIDIAN PIT
    36S   01W   26
    42-24-44N   122-47-34W   STONE
MERIDIAN SUB
    37S   01E   07   SW
    42-21-44N   122-45-03W   STONE(BASALT)
MIDDLE FORK ROGUE PIT
    37S   02W   25
    42-19-15N   122-52-50W   STONE

MIDNIGHT
    33S   01E   33   "NE,NW"
    42-39-54N   122-42-38W   BASALTS AND
    TUFF BRECCIAS   93 02
MILITARY GRAVEL PIT
    36S   02W   01
    42-27-49N   122-52-45W
    SAND & GRAVEL
MILLER
    37S   04W   19   "NW,NE"
    42-20-37N   123-12-56W
    AU   ARGILLITE   80 12
MILLIONAIRE
    36S   02W   36   "NE,NE,NE"
    42-25-36N   122-59-18W
    "META-ANDESITE, ARGILLITE, LIMESTONE"

MOCKING BIRD CHROME
     37S   03W   29
     42-19-14N   123-04-38W   CR
MOCKS GULCH PROSPECT
     40S   04W   17   "SW,NE"
     42-05-35N   123-11-58W
     METAVOLCANICS   81 02
MODOC PIT
     36S   02W   05
     42-28-32N   122-57-15W   SAND & GRAVEL
MODOC ROCK
     35S   01W   30   NW
     42-30-01N   122-52-16W   STONE
MOLYBDENUM PROSPECT
     36S   04W   16   "CENTER,N"
     42-26-43N   123-10-43W
     METAVOLCANICS   81 01
MORRIS PROPERTY
     36S   03W   25   S
     42-24-25N   123-00-22W
     SERPENTINE   80 12
MORTON GRANITE
37S   02W   19
     42-20-13N   122-58-07W   STONE(GRANITE)
MOSES & COLLINS CLAIMS
     40S   04W   35   SW
     42-02-41N   123-08-40W
     AU   GREENSTONE 81 01
MOUNTAIN KING
     34S   03W   36   "NE,SE"
     42-34-09N   122-59-40W   HG
     "AMPHIBOLITE, MICA SCHISTS"   80 10
MOUNTAIN QUEEN
     40S   04W   35   "SE,SW"
     42-02-33N   123-08-20W   DIORITE

MOUNTAIN VIEW
     33S   04W   17   "S,S,S"
     42-41-41N   123-12-02W   METABASALT
AND SERPENTINE   81 01
MOUNTAIN VIEW
     34S   04W   17
     42-36-34N   123-11-33W
MOUNTAIN VIEW
     37S   03W   06   SE
     42-22-39N   123-05-35W
     METAVOLCANICS   81 01

MT. PITT VIEW
      39S   01E   27
      42-09-11N   122-41-35W
      AU   GRANODIORITE
MURPHY PROSPECT
      40S   04W   09   SW
      42-06-12N   123-11-19W
      METAVOLCANIC ROCKS
NEATHAMER
      33S   04W   28
      42-40-37N   123-11-09W
      AU   STREAM GRAVEL   81 05
NEIL ROCK
      35S   02W   08   "NE,NE,SW"
      42-32-30N   122-57-58W
      "SANDSTONE; VERY COARSE, TUFFACEOUS, LITHIC"
          90 12
NELLIE GRAY GROUP PLACER
      40S   04W   21
      42-04-59N   123-11-13W
NELLIE WRIGHT
      36S   03W   24   SW
      42-25-30N   123-00-28W
      QUARTZ DIORITE/ANDESITE   81 01
NEW HOPE CLAIM
      40S   01W   20
      42-04-59N   122-50-36W   CR
      METASERPENTINITE   80 12
NEWSTROM RANCH
      36S   02E   34
      42-23-43N   122-34-31W
      RED VESICULAR TUFF 80 12
NICHOLS PROSPECT
      36S   02E   04
      42-28-19N   122-36-12W
      "TUFF, BASALT"   80 12
NONE SUCH
      38S   04W   23   CENTER
      42-15-09N   123-08-22W
      AU AG GREENSTONE 81 01
NORLING MINE
      37S   03W   26   "SW,SW"
      42-19-07N   123-01-59W
      AU AG GREENSTONE (ALTERED LAVAS)   81 01
NORTH CENTRAL CLAIM
      40S   03W   31   SE
      42-02-37N   123-05-57W
      SCHIST 81 01

NORTH POLE
        36S   04W   21
        42-25-40N   123-10-36W   AU
OFFENBACKER
        38S   03W   30
        42-13-57N   123-05-32W   AU
OLD FORT LANE
        36S   02W   19   "CENTER,N"
        42-25-45N   122-58-53W
        GRANODIORITE   81 01
ONION SPRINGS #1
        33S   04W   17
        42-42-23N   123-12-03W
        CR   SERPENTINE DUNITE
OPP
        37S   03W   36   NW
        42-18-43N   122-59-49W
        AU AG CU PB   "SILICEOUS AND CARBONACEOUS
        ARGILLITES, META-ANDESI"   80 12
OREGON BELLE MINE
        38S   03W   06   S
        42-17-25N   123-06-02W
        AU   GREENSTONE ARGILLITE ARGILLITE
OREGON GRANITE
        37S   02W   19   S
        42-20-13N   122-59-08W
        STONE(GRANITE)   QUARTZ DIORITE   81 01
OWL HOLLOW PROSPECT
        36S   04W   32   SW
        42-23-33N   123-12-06W
        AU   GREENSTONE 80 12

PALMER CREEK PLACERS
        40S   04W   01
        42-06-56N   123-07-20W
        AU   CHANNEL GRAVELS   81 05
PARKER BAR
        36S   02W   17
        42-26-00N   122-58-20W   SAND & GRAVEL
PARKER PIT
        36S   02W   20
        42-25-41N   122-57-31W   SAND & GRAVEL
PATRICK MANGANESE
        39S   01W   34
        42-08-22N   122-48-05W

PERKEYPILE MINE
 37S   03W   05   W
 42-22-55N   123-05-06W   81 01
PETERS MANGANESE
 39S   01W   17   "NW,NE"
 42-11-01N   122-50-29W
 SILICEOUS METASEDIMENTS 80 12
PHELPS PIT
 38S   02W   23   "NE,SE"
 42-14-58N   122-53-55W   STONE

PHILLIPS
 38S   01W   36
 42-13-19N   122-45-35W
 "METAVOLCANICS, METADIABASE"   80 10
PHOENIX GRAVEL PIT
 38S   01W   15
 42-16-03N   122-48-16W   SAND & GRAVEL
PILGRIM CLAIM
 39S   01W   14   SW
 42-10-36N   122-47-11W   SCHIST
PILOT ROCK
 40S   02E   34
 42-02-51N   122-34-21W   STONE
PITT VIEW
 33S   01W   22
 42-40-57N   122-48-35W
 ANDESITE FLOWS AND AGGLOMERATES
PLEASANT #1 & 2
 33S   04W   33
 42-39-36N   123-10-54W
 CR   SERPENTINE
PLEASANT CREEK PLACERS
 34S   04W   15
 42-36-59N   123-09-31W
 AU   COLLUVIUM AND CHANNEL GRAVELS.
POOLE AND PENCE PROSPECTS
 33S   01W   25   "CENTER,S"
 42-39-51N   122-45-41W   BASALT
PORCUPINE PLACER
 34S   04W   22
 42-36-33N   123-09-47W   AU
PRATER DEPOSIT
 39S   01W   30
 42-08-44N   122-51-13W   CR   DUNITE
PURVIS PROSPECT
 41S   02W   10   SE
 42-00-48N   122-55-09W   QUARTZ-MICA SCHIST

QUARTZ MOUNTAIN
    30S   02E   34   "N,N"
    42-55-18N   122-34-12W
    SILICA(LODE) RHYOLITE   81 01
QUARTZ-SERPENTINE PIT
    35S   03W   29   SE
    42-29-34N   123-04-50W
    QUARTZ & SERPENTINE
QUEEN ANNE MINE
    39S   02W   09   CENTER
    42-11-32N   122-56-26W   AU   ARGILLITE
QUEEN MARY PROSPECT
    39S   01W   27
    42-09-20N   122-47-47W
QUEENS BRANCH QUARRY
    35S   04W   10
    42-32-14N   123-09-16W   STONE
R C GILBERT
    37S   02W   19
    42-20-17N   122-58-40W   STONE
RAINBOW CHROME
    33S   04W   18
    42-42-06N   123-13-40W

RAMSEY CANYON QUARRY
    35S   02W   06   NW
    42-33-39N   122-59-30W   STONE

RAMSEY LAKE
    35S   02W   18   "NW,SW"
    42-31-30N   122-59-21W
    AU   "QUARTZ MICA SCHIST, QUASRTZITE"
    90 12
RASPBERRY CREEK
    34S   03W   15   "NE,SW"
    42-36-48N   123-02-45W
    ASBESTOS   SERPENTINE   81 01
RATTLESNAKE MINE
    37S   03W   05   SW
    42-22-47N   123-05-00W   GREENSTONE
    81 01
RAY MINE
    37S   03W   05   "S,NW"
    42-22-56N   123-05-06W
    AU   GREENSTONE 81 01

RAYOME PROSPECT
        33S   01E   33   "NW,NW"
        42-39-42N   122-39-42W
        BASALTIC LAVAS AND BRECCIAS   93 02
RED BEAN CLAIM PLACER
        40S   04W   06   NE
        42-07-28N   123-12-38W   AU
RED BELL
        38S   03W   24
        42-14-52N   123-00-23W   AU
RED CIRCLE
        40S   04W   23
        42-04-30N   123-08-36W   AU

RED FERN MINE
        36S   04W   17   "SE,SE,NE"
        42-26-00N   123-11-27W
        GREENSTONE 81 01
RED GOLD
        40S   01E   25
        42-03-46N   122-38-50W
RED MOUNTAIN MINES
        40S   01W   29   SE
        42-03-43N   122-50-13W
        CR   "DUNITE, METAPERIDOTITE" 81 01
RED OAK
        36S   03W   04
        42-28-23N   123-04-10W
RED STAR PROSPECT
        41S   02W   17   "W,NW"
        42-00-35N   122-58-09W
        CHLORITE-EPIDOTE SCHIST   81 02
REED
        35S   03W   01   E
        42-33-25N   122-59-42W
        AU   GREENSTONE 77 05
REEDER
        39S   01E   20   SE
        42-09-29N   122-43-11W
        AU   QUARTZ DIORITE   80 12
REESE CREEK QUARRY
        35S   01E   07
        42-32-30N   122-44-29W
        SAND & GRAVEL

REVENUE POCKET
        37S   03W   11   SE
        42-21-53N   123-00-57W
        AU AG "ARGILLITE, LIMESTONE"
ROBCO INC
        36S   03W   18
        42-26-09N   123-05-38W
        SAND & GRAVEL
ROGER'S CINNABAR PROPERTIES
        33S   01W   25   SE
        42-38-45N   122-45-37W
ROGERS' TILLER-TRAIL
        32S   01W   19
        42-46-00N   122-52-20W
ROGUE AGGREGATES PIT
        37S   02W   25
        42-19-20N   122-53-09W
        SAND & GRAVEL
ROGUE RIVER PROSPECT
        33S   01E   33   "SW,SW"
        42-39-13N   122-42-53W
        BASALT (?) FLOWS   93 02
ROOSEVELT GROUP
        40S   04W   21
        42-04-31N   123-10-31W
        AU   STREAM GRAVEL   81 02

ROXANA
        34S   02W   05   E
        42-38-59N   122-57-48W
        HG   SCHISTOSE GREENSTONE (ANDESITE AND
        BASALT LAVAS AN   81 02

ROXY ANN
        37S   01W   14   SE
        42-20-49N   122-46-41W
        STONE(BASALT)
ROXY ANNE
        37S   01W   14   "W,NE"
        42-21-22N   122-48-15W
        "SANDSTONE, SHALE" 81 01
RUBY MINES
        41S   01W   05
        42-02-12N   122-50-41W
RUBY QUICKSILVER PROSPECT
        40S   03W   34   SE
        42-02-34N   123-01-57W
        HG   SERPENTINITE GRANODIORITE   81 02

"RYAN,CHENEY, & BATEN"
    36S   01E   04
    42-27-43N   122-42-22W   STONE
RYNO
    34S   01E   03
    42-38-35N   122-41-11W
SAFEWAY
    36S   03W   01
    42-28-16N   123-00-42W   AU
SARDINE CREEK PLACERS
    36S   03W   05   "S,SE"
    42-27-48N   123-04-25W
    AU   CHANNEL GRAVELS   81 05
SAYLER SITE
    37S   02W   19   "W,CENTER"
    42-20-17N   122-59-34W
    STONE(GRANITE)
SCHAFFER CLAIM
    36S   03W   24   S
    42-25-23N   123-00-13W
    TONALITE   81 01
SCHMIDT PROSPECT
    37S   03W   05   "NE,NW"
    42-23-16N   123-04-56W
    METAVOLCANICS   81 01
SCHNACK #1 & 2
    39S   02W   35
    42-08-17N   123-01-02W
SCHNACK BROS. 1 & 2
    39S   01W   35   "SE,NE"
    42-08-15N   122-46-56W
    QUARTZ DIORITE   81 01
SCOTT MANGANESE PROSPECT
    37S   02W   01
    42-19-36N   123-01-14W
SCOTT MINE
    38S   04W   13   "NW,SW"'
    42-15-48N   123-07-48W
    METAVOLCANICS   81 01
SEAMAN BAR
    36S   04W   20
    42-25-09N   123-11-53W   AU
SEATTLE BAR MARBLE
    41S   04W   02   SW
    42-00-59N   123-08-59W   STONE(LIMESTONE)
    METASEDIMENTS

SEVEN L. MOLY PROSPECT
40S   01E   27
42-03-32N   122-41-04W
QUARTZ DIORITE
SEVENTY THREE
35S   03W   01
42-33-46N   123-00-06W
SCHIST 81 02
SHADY COVE/LONG BRANCH
34S   01W   18
42-37-06N   122-51-30W
STONE
SHAMROCK
34S   02W   19   E
42-36-04N   122-58-32W
CU NI   QUARTZ-MICA SCHIST 79 04
SHAMROCK MANGANESE
34S   02W   19
42-35-51N   122-58-32W
MN   QUARTZITE SCHIST SCHIST   81 01
SHANDA MINE
39S   02W   05   "E,E"
42-12-26N   122-57-23W
METAVOLCANICS   81 01
SHARON LAKE QUARRY
38S   03E   18
42-15-35N   122-30-20W
STONE
SHORTY HOPE
39S   01W   12   NW
42-12-13N   122-46-14W
AU AG HORNFEISED METAVOLCANICS   80 12

SHULL
34S   02W   13
42-36-55N   122-53-05W
VOLCANIC ROCK   81 02
SILVER OCCURRENCE
38S   02E   30
42-13-52N   122-38-00W
ANDESITE BRECCIA
SINNIGER
36S   03W   28
42-24-43N   123-03-34W   AU
SISKIYOU GAP PROSPECT
40S   01W   34   NE
42-03-09N   122-48-03W
HORNBLENDE QUARTZ DIORITE   81 02

297

SKYLINE
      39S   01E   30   NW
      42-09-15N   122-44-45W
      AU AG GRANODIORITE   80 12
SMUGGLER PROSPECT
      36S   03W   02   NW
      42-28-19N   123-01-35W
      AU AG ARGILLITE   81 01
SNAPSHOT CLAIM
      39S   01W   23   "SW,SW,SW"
      42-09-29N   122-47-41W
      TONALITE   81 01
SNAVELY ROCK
      39S   03W   13
      42-10-59N   123-00-33W   STONE
SNOWY RIDGE
      41S   02W   14
      42-00-13N   122-53-41W
      CR   "METASERPENTINITES, DUNITE"
SODA CREEK
      36S   02W   17
      42-26-00N   122-57-37W   SAND & GRAVEL
SPARKS
      35S   04W   02   E
      42-33-26N   123-07-56W
      PEGMATITE DIORITE 81 01
SQUAW CREEK COPPER
      41S   03W   05   NW
      42-02-11N   123-05-15W
      CHLORITE SCHIST   81 01
STANDING BUCK #1-2
      36S   03W   19
      42-25-30N   123-06-02W   AU
STANLEY AND BROWN PROSPECT
      36S   01E   11   "E,NW"
      42-27-30N   122-40-09W
      HG   ANDESITIC FLOWS AND PYROCLASTICS
STAPLES BAR
      38S   03W   28
      42-14-12N   123-04-05W
      SAND & GRAVEL
STAR F. RANCH
      36S   02E   11
      42-27-15N   122-32-54W
      RED TUFF   80 12
STAR MINE
      39S   04W   06   NE
      42-13-41N   123-12-57W
      AU AG "ARGILLITE, METAVOLCANICS"   80 12

STARR PROSPECT
>   34S   04W   21
>   42-36-18N   123-10-54W
>   SAND STONE   81 01

STEAMBOAT
>   40S   04W   20   "NE,NE"
>   42-04-55N   123-11-48W   AU

STEAMBOAT CINNABAR
>   40S   04W   18   "NW,NE"
>   42-05-51N   123-12-55W
>   ARGILLITE AND SANDSTONE 81 02

STEARNS MOLY PROSPECT
>   36S   03W   14   "SW,NE"
>   42-26-30N   123-01-06W   METAGABBRO
>   81 01

STERLING GOLD QUARTZ MINING CO.
>   38S   02W   33   "N,N,N"
>   42-13-43N   122-56-35W
>   AU AG ANDESITE   81 01

STEWART
>   40S   03W   30
>   42-03-34N   123-06-35W

STROMBERG SILICA
>   35S   04W   27   "CENTER,S"
>   42-29-41N   123-09-34W
>   QUARTZ DIORITE   81 01

SUGAR PINE
>   38S   02W   07   SE
>   42-16-27N   122-58-34W

SUMMIT LAKE
>   41S   03W   11   CENTER
>   42-00-51N   123-01-38W
>   TALC(SOAPSTONE) "SERPENTINE, QUARTZ-MICA
>   AND EPIDOTE-ACTINOLITE SCH"

SUNDOWN PROSPECT
>   38S   03W   07   "NE,NW"
>   42-17-07N   123-06-17W   81 01

SUNRISE
>   38S   02W   23   "NW,NW"
>   42-15-22N   122-54-37W   81 01

SUNRISE PROSPECT
>   38S   04W   06   "NE,SE"
>   42-17-33N   123-12-34W
>   DIORITE   81 01

SUNSET MINE
>   34S   04W   03
>   42-38-20N   123-09-11W

SWACKER FLAT
    37S   04W   12   NE
    42-22-18N   123-07-24W
    AU   FOOTS CREEK GRAVEL   81 02
SYKES CREEK MINING CO.
    35S   04W   01
    42-33-26N   123-07-19W
SYLVANITE MINE
    36S   03W   02   S
    42-27-48N   123-01-19W
    AU AG PB W   "ARGILLITE, METAVOLCANICS"
    80 12
T&M ANTIMONY
    32S   01W   11
    42-45-35N   122-47-02W
TABLE ROCK SITE
    36S   02W   15
    42-26-42N   122-55-50W
    SAND & GRAVEL
TAYLOR
    39S   01W   14
    42-10-46N   122-47-06W
    QUARTZITE   81 01
TELKAMP PLACER
    34S   04W   21
    42-35-41N   123-10-15W   AU
THE OLD STURGIS
    38S   03W   10
    42-16-50N   123-02-48W   AU
THE STAR
    37S   02W   05
    42-24-44N   122-57-49W   AU
THOMPSON CREEK PLACERS
    38S   04W   28   N
    42-14-09N   123-10-54W
    AU   CHANNEL GRAVELS   81 05
THORNDIKE PIT
    37S   02E   36   "SE,SE"
    42-18-16N   122-31-27W
    STONE
THRUSH
    33S   04W   16   N
    42-42-30N   123-10-40W
    SERPENTINE   81 01
TINPAN
    36S   03W   31   SE
    42-23-30N   123-05-32W
    "ARGILLITE, LIMESTONE, META-ANDESITE"

TITO PROSPECT
        37S   04W   14   "N,NW"
        42-21-38N   123-08-40W
        QUARTZ DIORITE   81 01
TOLMAN IRON
        36S   03W   03   SW
        42-27-56N   123-02-39W
        "SCHIST, ARGILLITE, LIMESTONE, QUARTZITE"
            81 01
TOPROCK
        34S   01W   30   NE
        42-35-18N   122-51-51W   STONE

TOWER ROCK PROSPECT
        34S   02W   02   "NE,NE"
        42-38-49N   122-53-42W
        TUFF AND RHYOLITE BRECCIA   81 05
TOWN MINE
        37S   03W   25   S
        42-19-20N   123-00-20W
        AU AG CU   ARGILLITE   81 01
TRAIL URANIUM
        33S   01W   31   "SW,SW,SW"
        42-39-07N   122-52-25W
        RHYOLITIC FLOWS AND TUFFS   81 05
TREE MINE
        40S   03W   05   SE
        42-06-59N   123-04-37W
        METASEDIMENTS   81 01
TRUST BUSTER
        35S   03W   36   "W,W"
        42-29-09N   123-00-42W
        AU AG METASEDIMENTS

TUCKER/GILBERT
        36S   01W   04
        42-28-15N   122-49-11W   STONE
TUER PIT
        36S   02W   28
        42-24-38N   122-56-24W   SAND & GRAVEL
TYRRELL MINE
        37S   02E   10   "NW,NW"
        42-22-26N   122-34-41W
        MN   RED TUFF   80 12
TYSON PLACER
        40S   04W   17
        42-05-21N   123-11-49W

UPPER APPLEGATE PLACERS
      40S   04W   19   "NW,NE"
      42-04-52N   123-12-59W
      AU   STREAM GRAVELS   81 04
US CINDER PIT
      35S   04W   36
      42-29-00N   123-07-11W
      CINDERS
VAN CURLER
      39S   01E   06
      42-12-28N   122-44-40W
      AU   GRANODIORITE   81 01
VESTAL GROUP
      35S   01E   07   SW
      42-32-40N   122-43-14W
           ANDESITIC TUFF   80 12
VICTOR PROSPECT
      38S   04W   10   "SE,SE,NE"
      42-16-23N   123-09-06W
      METAVOLCANICS   81 02
VICTORY GROUP
      36S   03W   12   SW
      42-27-06N   123-00-39W   AMPHIBOLITE
      81 02
VROMAN PLACER
      36S   03W   04
      42-28-13N   123-04-10W   AU
WAGNER PLACER
      40S   04W   20
      42-04-29N   123-12-31W   AU

WAGON TRAIL QUARRY
      37S   03W   36
      42-18-19N   123-00-25W   STONE
WALKER CREEK
      37S   02W   19   SW
      42-20-03N   122-59-21W   STONE(GRANITE)
WALKER CREEK
      37S   02W   19   "NE,SW"
      42-20-13N   122-59-07W   STONE(GRANITE)
WAR EAGLE
      34S   02W   17   "NW,NW"
      42-37-15N   122-57-42W   HG   AMPHIBOLITE
      SCHIST BIOTITE SCHIST
WAR EAGLE (COAL AREA)
      34S   02W   16   "NW,NW"
      42-37-12N   122-57-13W
      LIGNITIC COAL AND CARBONACEOUS SILTSTONE

WARD CREEK
     36S   04W   01
     42-27-53N   123-07-07W
WARD CREEK MANGANESE
     35S   04W   36   "S,SW"
     42-28-41N   123-06-43W
     SHALE SANDSTONE   81 02
WARD CREEK PLACERS
     35S   04W   36
     42-28-44N   123-06-56W
     AU   CHANNEL GRAVELS   81 05
WARNER
     33S   04W   04   NE
     42-44-12N   123-10-21W
     AU AG METAGABBRO 81 02
WARNER ROCK
     36S   04W   22
     42-25-54N   123-09-28W   STONE
WATER GULCH CLAIM
     39S   01W   29
     42-08-38N   122-50-57W   METADUNITE
WELL NICKEL PROSPECT
     40S   04W   28
     42-03-44N   123-11-50W
WELLMAN
     37S   02W   08
     42-22-07N   122-57-31W   STONE
WELLS PIT
     38S   03W   01   NE
     42-17-45N   122-59-55W   STONE
WELLS PROSPECT
     40S   04W   33   NE
     42-03-02N   123-11-14W   80 12
WEST EVANS CREEK COPPER
     34S   03W   10
     42-37-22N   123-02-27W   METAGABBRO
WESTSIDE CARBERRY CREEK
     40S   04W   33
     42-03-05N   123-10-29W   SERPENTINE
WET BRUSH PROSPECT
     33S   03W   12
     42-43-15N   122-59-38W
     CHLORITE SCHISTS
WEYERHAUSER CINDER PIT
     39S   04E   35
     42-08-01N   122-18-53W
     CINDERS

WHITE HORSE MINING CO.
        36S   03W   03   SW
        42-27-52N   123-02-55W
        AU   METASEDIMENT   81 02
WHITE OAK PROSPECT
        38S   04W   06   "S,SE"
        42-17-18N   123-12-48W
        DIORITE   81 02
WHITNEY
        36S   03W   13   "NE,SW"
        42-27-05N   123-00-49W
        AU   PYROXENITE HORNFELS HORNFELS

WHITTLE
        36S   02W   16
        42-26-36N   122-56-22W
        SAND & GRAVEL
WHITTLE BAR
        36S   02W   15
        42-26-42N   122-55-50W
        SAND & GRAVEL
WHITTLE BAR & PLANT
        37S   02W   25
        42-19-15N   122-52-50W
        SAND & GRAVEL
WILLIAMS
        34S   04W   32   NE
        42-34-34N   123-11-46W
        AU   GRAVEL   81 05
WILLIAMS DREDGE
        38S   02W   30   "SW,NW"
        42-14-12N   122-59-12W
        AU   STREAM GRAVELS
WILLIAMSON MINE
        37S   03W   25   "SW,NW"
        42-19-06N   123-00-39W
        AU   METAVOLCANICS   81 02

WILLOW CREEK/DEE EARL IDE TRUCKING
        37S   02W   06
        42-22-455   122-58-455   STONE
WILLOW SPRINGS
        36S   03W   30
        42-24-44N   123-05-59W   AU
WILSON GEORGE JENNIE
        36S   03W   03
        42-28-06N   123-02-04W

WILTON WHITE GRAVEL PIT
      36S   02W   16
      42-26-00N   122-57-04W
      SAND & GRAVEL
WINCHESTER-HOUSTEN
      37S   02W   06
      42-23-10N   122-58-49W
      STREAM GRAVEL
WOODPECKER GROUP
      40S   04W   16   SW
      42-05-01N   123-11-23W
      METAVOLCANIC ROCKS   81 02
WOODS PROSPECT
      35S   04W   25
      42-30-17N   123-07-36W
      PHYLLITE   81 02
WOOLF TUNGSTEN
      37S   04W   14   "NW,SW"
      42-21-09N   123-08-54W
      METAVOLCANIC
WOUNDED BUCK
      39S   03W   34
      42-08-12N   123-02-28W   AU
WRIGHT MINE
      38S   04W   23   "N,NW"
      42-15-27N   123-08-39W
      GREENSTONE
YELLOW KING MINE
      37S   03W   26   NE
      42-19-32N   123-01-01W
      META-ANDESITE
YELLOWJACKET
      39S   04W   35   "N,N"
      42-07-43N   123-08-58W
      METAVOLCANIC

Glossary of Terms

Air fall Pumice: Porous rock created when violent volcanic explosions hurl molten rock into the air, creating very light rock. Pumice is so full of space and so lightweight it will actually float in water.

Andesite Vein: A gray, fine grained volcanic rock, chiefly plagioclase and feldspar.

Antimony: A metallic element having four allotropic forms the most common of which is a hard, extremely brittle, lustrous, silver-white, crystalline material. It is used in a wide variety of alloys, especially with lead battery plates, and in semiconductor devices, and ceramic products.

Assay: The qualitative or quantitative analysis of a substance. Especially of an ore or drug.

Basalt: A hard, dense, dark volcanic rock composed chiefly of plagioclase, augite, and magnetite, often having a glassy appearance.

Chlorite: A generally green or black secondary mineral, often formed of dark rock minerals.

Cinnabar: A heavy reddish mercuric sulfide, HgS, that is the principal ore of mercury. Also called "vermillion".

Copper: A ductile, malleable, reddish-brown metallic element that is an excellent conductor of heat and electricity and is widely used for electrical wiring, water piping, and corrosion-resistant parts either pure or in alloys such as brass and bronze.

Drainage: The action of or a given method of draining.

Dredge: Any of various machines equipped with scooping or sucking devices used in deepening harbors and waterways and underwater mining.

Flow Rocks: Rocks that have been forced down any given stream or vein from their originating source.

Gabbro: A usually course-grained igneous rock composed chiefly of igneous rock composed chiefly of calcic plagioclase and pyroxene, sometimes with other minerals. Also called "norite". Any course grained igneous rock.

Garnierite: An earthy, apple-green mineral, an important nickel ore.

Geomorphic: Of, relating to, or resembling the earth, its shape, or surface configuration.

Gold: A soft, yellowish, corrosion resistant element, the most malleable and ductile metal, occurring in veins and alluvial deposits, and recovered by mining or by panning or sluicing. It is a good thermal and electrical conductor, is generally alloyed to increase it's strength , and is used as an international monetary standard, in jewelry, for decoration, and as a plated coating on a wide variety of electrical and mechanical components. Something regarded as having a certain wealth or goodness: A heart of Gold.

Greenstone: Any of various altered basic igneous rocks colored green by chlorite, hornblende, and epidote.

Hydraulic mining: Of, operated, involving, or moved by a fluid.

Intruded: To thrust (molten rock) into a stratum.

Lode: 1. mining. A. A fissure in a rock formation that is filled with a metalliferous ore.

Long Tom Sluice box: A sluice box with an average length of 6' to 100' or more.

Mesozoic age: Of, belonging to , or designating the third era of geologic time. Includes the Cretaceous, Jurassic, and Triassic periods, and is characterized by the predominance of reptilian life forms.

Metamorphosed: To cause change in form, structure, or character.

Mine: An excavation of earth in order to release free metals.

Mineral resource: An ore. Of, or pertaining to minerals.

Nickel: A silvery, hard, ductile, ferro-magnetic metallic element.

Open Claim: A mineral rights claim is available for anyone to stake claim on. Anyone is able to stake claim on an open claim as long as they follow the laws and regulations set forth by the local municipality.

Placer: A glacial or alluvial deposit of sand or gravel containing eroded particles of valuable materials.

Placer Mining: The obtaining of minerals from placers by washing or dredging.

Platinum: A silver-white metallic material occurring world wide, usually mixed with other metals, such as iridium, osmium, or nickel. It is ductile and malleable, does not oxidize in air, and is used in electrical components, jewelry, electroplating, and dentistry, and as a catalyst.

Pleistocene: Of, or relating to the desxignating the geologic time, rock series, and sedimentary deposits of the earlier of the two epochs of the quarternary period, characterized by the alternate appearance and recession of the northern glacian and the appearance of the progenitors of man.

Pliocene: Of, relating to, or designating the geologic time, rock series, and sedimentary deposits of the last of the five epochs of the Tertiary period, characterized by the appearances of distinctly modern plants and animals.

Quartz: A hard, crystalline, vitreous mineral silicone dioxide, SiO_2, found world-wide as a component of sandstone and granite , or as pure crystals in such varieties as agate, chalcedony, chert, flint, opel, and rock crystal.

Sandstone: Variously colored sedimentary rock composed predominantly of sand like quartz grains cemented by lime, silica, or other materials.

Sediment: Material that settles to the bottom of a liquid; dregs; lees.

Serpentine: A greenish, brownish, spotted material.

Shale: A fissile rock composed of laminated layers of claylike, fine-grained sediments.

Siltstone: A sedimentary material consisting of fine mineral particles intermediate in size between sand and clay.

Strata: A horizontal layer of any material; especially, one of several parallel layers arranged one on top of another.

Tertiary: Of, or relating to the geologic time, system of rocks and sedimentary of the first period of the cenozoic, extending from the cretaceous period of the Mesozoic era to the quaternary period of the Cenozoic era, characterized by the appearance of modern flora and of apes and other large mammals.

Tin: A malleable , silvery metallic element obtained chiefly from cassiterite. It is used to coat other metals to prevent corrosion, and forms part of numerous alloys, such as soft solder, pewter, type metal, and bronze.

Topographic: The detailed and accurate description of a place or region.

Tributary: A stream or river flowing into another stream or river.

Volcanic Rocks: Produced by, or discharged from a Volcano.

Pictured here is the Gold Bug Mines on the Rouge River. Very easy to find in fact it's so easy I am sitting in my car taking the picture.

Notes and Information

www.ingramcontent.com/pod-product-compliance
Lightning Source LLC
Chambersburg PA
CBHW071358170526
45165CB00001B/97